Assessment of the Proliferation of Certain Remotely Piloted Aircraft Systems

Response to Section 1276 of the National Defense Authorization Act for Fiscal Year 2017

George Nacouzi, J.D. Williams, Brian Dolan, Anne Stickells, David Luckey,

Colin Ludwig, Jia Xu, Yuliya Shokh, Daniel M. Gerstein, Michael H. Decker

Prepared for the Deputy Director for Battlespace Awareness (J28), Joint Chiefs of Staff

For more information on this publication, visit www.rand.org/t/RR2369

Library of Congress Cataloging-in-Publication Data is available for this publication.

ISBN: 978-1-9774-0034-5

Published by the RAND Corporation, Santa Monica, Calif.

© Copyright 2018 RAND Corporation

RAND® is a registered trademark.

Cover Image by Lt. Col. Leslie Pratt, U.S. Air Force.

Support RAND

Make a tax-deductible charitable contribution at

www.rand.org/giving/contribute

www.rand.org

Preface

The National Defense Authorization Act (NDAA) for Fiscal Year 2017, Section 1276, requires an independent assessment directed by the Chairman of the Joint Chiefs of Staff to report on the "impact to United States national security interests of the proliferation of remotely piloted aircraft that are assessed to be 'Category I' items under the Missile Technology Control Regime" (Pub. L. No. 114-328, 2016). The NDAA requires this evaluation, in the form of a report, to be delivered to the congressional defense committees. The NDAA requires several specific assessments, including the threat posed to the United States, impact on allies and partners, and the benefits and risks of continuing to limit the export of these aircraft. To respond to the NDAA, RAND Corporation researchers conducted literature reviews, collected and analyzed publicly available and classified data and information, and conducted interviews with subject-matter experts.

This research was sponsored by the Office of the Chairman of the Joint Chiefs of Staff Directorate for Intelligence and conducted within the Cyber and Intelligence Policy Center of the RAND National Defense Research Institute, a federally funded research and development center sponsored by the Office of the Secretary of Defense, the Joint Staff, the Unified Combatant Commands, the Navy, the Marine Corps, the defense agencies, and the defense Intelligence Community.

For more information on the RAND Cyber and Intelligence Policy Center, see http://www.rand.org/nsrd/ndri/centers/intel.html or contact the director (contact information is provided on the webpage).

Contents

Figures and Tables

Figures

Tables

Summary

Section 1276 of the National Defense Authorization Act (NDAA) for Fiscal Year 2017 requires an independent assessment, directed by the Chairman of the Joint Chiefs of Staff, of the impact that certain remotely piloted aircraft (RPA) governed by the Missile Technology Control Regime (MTCR) have on U.S. national security interests (Pub. L. No. 114-328, 2016). The NDAA requires that this evaluation, in the form of a report, be delivered to the congressional defense committees. The congressional language specifically requires that the assessment include evaluation in six areas:

> (1) A qualitative and quantitative assessment of the scope and scale of the proliferation of remotely piloted aircraft that are "Category I" items under the Missile Technology Control Regime. (2) An assessment of the threat posed to United States interests as a result of the proliferation of such aircraft to adversaries. (3) An assessment of the impact of the proliferation of such aircraft on the combat capabilities of and interoperability with allies and partners of the United States. (4) An analysis of the degree to which the United States has limited the proliferation of such aircraft as a result of the application of a "strong presumption of denial" for exports of such aircraft. (5) An assessment of the benefits and risks of continuing to limit exports of such aircraft. (6) Such other matters as the Chairman considers appropriate. (Pub. L. No. 114-328, 2016)

To respond to the NDAA, at the request of the Joint Chiefs of Staff, RAND Corporation researchers conducted literature reviews, collected and analyzed publicly available and classified data and information, and met with subject-matter experts to consider their expert feedback. This report contains the results of our findings and assessment. It provides the assessments with respect to category I items (subject to a strong presumption of denial for export) as the NDAA requires and, as discussed in the next section, includes an assessment of a subset of category II items (exportable subject to certain licensing requirements and legislated guidelines) where the implications of the proliferation of such items have particular relevance to the first three requests of the NDAA study.[1]

Remotely Piloted Aircraft and the Missile Technology Control Regime

This report applies to unmanned or uninhabited aircraft. These vehicles have various designations throughout the U.S. government and industry, including *RPA, unmanned aerial vehi-*

---

[1] The next section provides more detail on MTCR categories I and II.

cles (UAVs), and *unmanned aircraft*. The MTCR defines UAVs differently to include not only unmanned or uninhabited aircraft but also reconnaissance drones, target drones, and cruise missiles. To avoid misinterpretation, we use the most common term, *UAV*, rather than *RPA*, to refer to unmanned or uninhabited aircraft, which are the subject of this report.[2] The selection and use of the term *UAV* have no effect on any of the assessments or findings in this report.

The MTCR was formed in 1987 as

> an informal and voluntary association of [35] countries which share the goals of non-proliferation of unmanned delivery systems capable of delivering weapons of mass destruction (other than manned aircraft), and which seek to coordinate national export licensing efforts aimed at preventing their proliferation. (MTCR, 2010, p. ii)

The MTCR export-control list has two categories of items:

- **Category I** items include any complete rocket or UAV capable of delivering a payload of at least 500 kg to a range of at least 300 km, its major complete subsystems, related software and technology, and specially designed production facilities for these items. Category I items are subject to a "strong presumption of denial" and are approved for export on rare occasions only (MTCR, undated [b]).
- **Category II** items include other less sensitive and dual-use components, as well as complete systems capable of a range of at least 300 km, regardless of payload. Their export is less restrictive than category I systems but are subject to licensing requirements, taking into consideration the nonproliferation factors specified in the MTCR guidelines (MTCR, undated [b]).

Informed by our early findings, we extended the types of UAVs to consider in this assessment beyond category I aircraft as the NDAA requests. For the assessment of the NDAA evaluation areas 1–3, we include a subset of somewhat smaller UAVs that we call *near–category I."* Although technically category II items, these systems have a payload capacity of between 300 and 500 kg (and a range greater than 300 km), perform many of the missions that category I systems do, and are significantly less expensive. Moreover, as sensors and even weapons become more miniaturized, the relative capabilities, and therefore importance, of near–category I platforms can be expected to grow and become an increasingly relevant threat. The inclusion of near–category I systems in the assessments is not meant to expand the definitions in the MTCR but rather to highlight that UAVs with payloads smaller than 500 kg can perform missions that are similar to those of many category I UAVs, as shown in Table S.1, so an assessment is needed of their proliferation, the threats they pose to the United States, and their impact on the capabilities of allies and partners of the United States. We use the term *large UAV* to refer to the combination of category I and near–category I systems.

---

[2]   Although *RPA* is a proper description of unmanned aircraft, it can have the connotation that a pilot is remotely flying the aircraft. Some UAVs can fly autonomously with minimal to no input from the controller.

Table S.1
Estimated Impacts of Missile Technology Control Regime Category I Export Controls

Impact	Positive	Somewhat Positive	Neutral	Somewhat Negative	Negative
Security		Protection of U.S. technology	Threat to U.S. forces	Interoperability; operational expertise	Allies' capabilities
Economic			Cost to U.S. customers	Future opportunities	Global market share; U.S. industrial base; research and development
Political	MTCR implications	Misuse of UAVs	Global proliferation	International motivation to develop UAVs	Partner relations

Assessment of Scope and Scale of the Proliferation of Unmanned Aerial Vehicles

For the purposes of this study, we assessed proliferation by measuring how many nations operate large UAVs, regardless of how many they have. Proliferation can occur through any of three principal methods: importing a complete system, converting an available system, or acquiring the capability to develop and manufacture large UAVs indigenously. Only about a dozen countries currently operate category I UAVs, which might be due to MTCR restrictions, mission needs, or unit cost. However, manufacturers have been skirting the limits of the MTCR by exporting long-range vehicles with payloads between 300 and 500 kg. These near–category I vehicles are quickly proliferating.

Figure S.1 captures the scope of category I and II UAV proliferation, while Figure S.2 illustrates their global spread. Approximately ten nations currently operate category I UAVs, while more than 15 nations operate near–category I UAVs. A small handful of countries have been responsible for the majority of large-UAV exports in recent years. Specifically, Israel and the United States have been the leading UAV exporters for many years, with the United States believed to be the sole exporter of category I UAV systems until 2015. Recently, China has become a significant exporter of near–category I UAVs, with the prospect of becoming a major exporter of category I systems in the near future. Other new exporters, such as the United Arab Emirates (UAE), are also expected to become major players, with the UAE already having exported category I Yabhon United 40 (Smart Eye 2) UAVs to Russia and claiming that it has received a large number of orders for the United 40 from unnamed recipients. We assess, based on current reporting, that the proliferation of large UAVs is accelerating. For example, China has agreed to set up a UAV manufacturing plant in Saudi Arabia for up to 300 new UAVs (possibly both category I and near–category I). China and the UAE are not only marketing large UAVs (including category I) but also offering to build factories for coproduction. These regional factories could further exacerbate the proliferation of large UAVs to the degree that these systems are exported to other nations.

Figure S.1
Scope of the Proliferation of Unmanned Aerial Vehicles

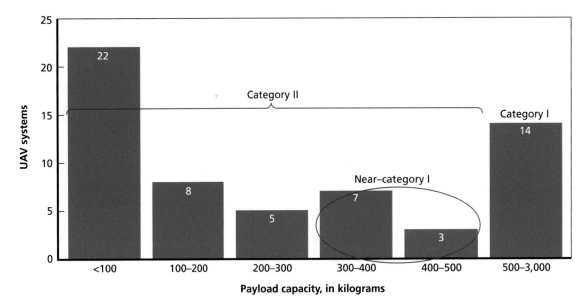

SOURCES: "All the World's Aircraft: Unmanned," undated; "Elbit Hermes 900 (Kochav) Medium Altitude, Long Endurance (MALE) Unmanned Aerial Vehicle (UAV)," 2017; Lappin, 2017; Piaggio Aerospace, undated; "Heron TP (Eitan) MALE UAV," undated; "Chang Hong-5 (CH-5) Combat and Reconnaissance Drone," undated; Aeronautics, undated.
NOTE: The figure shows UAVs with ranges greater than 300 km at these payload weights.
RAND RR2369-S.1

Figure S.2
Global Proliferation of Large Unmanned Aerial Vehicles

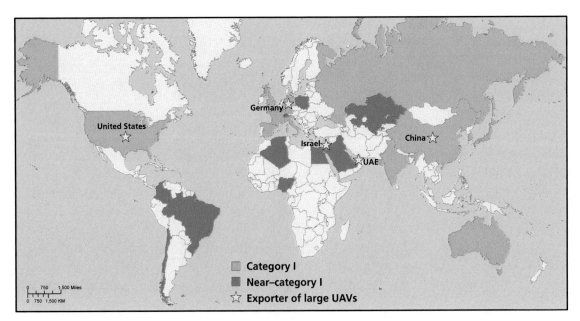

SOURCES: "All the World's Aircraft," undated; "Elbit Hermes 900 (Kochav) Medium Altitude, Long Endurance (MALE) Unmanned Aerial Vehicle (UAV)," 2017; "Heron TP (Eitan) MALE UAV," undated; "Chang Hong-5 (CH-5) Combat and Reconnaissance Drone," undated; Lappin, 2017; Piaggio Aerospace, undated; Aeronautics, undated.
RAND RR2369-S.2

The Threat That Proliferation of Large Unmanned Aerial Vehicles Poses to U.S. Interests

Regardless of U.S. or global policies concerning the export of UAVs of various sizes, we see these systems posing an incremental but growing threat to U.S. and allied military operations. Major potential adversaries—China, Russia, and Iran—recognize the utility of this capability and are producing many types of UAVs, including category I systems, for their armed forces (Lin and Singer, 2017a). Proliferation of these UAVs to potential adversary nations would, to some extent, complicate U.S. operations involving these nations. Given the continuing evolution and proliferation of both small and large UAVs, it is likely that, in future conflicts, U.S. forces will have to cope with adversaries equipped with different types and sizes of UAVs, both armed and unarmed.[3]

The primary threat from current category I (and near–category I) UAV systems is their ability to conduct intelligence, surveillance, and reconnaissance (ISR) operations against U.S. forces prior to hostilities. Adversaries that would otherwise have difficulty detecting U.S. force deployments, monitoring U.S. operations, and maintaining targeting data on U.S. units can employ UAVs to maintain situational awareness of U.S. capabilities. Although near–category I (and some smaller) UAV systems can perform this function, the larger payload of the category I systems enables larger, more-capable sensors that further increase the reach and endurance of UAVs conducting ISR. This capability is especially useful in monitoring ongoing force deployments in areas adjacent to coastlines or borders, where standoff surveillance is possible, and in the maritime domain. Additionally, for weaponized UAVs, the larger category I has a larger weapon capacity and could therefore have more-lethal effects.

During hostilities, battlefield surveillance, targeting, electronic warfare (e.g., jamming), and lethal engagement are all potential threats from UAVs. Large UAVs represent less of a threat in this environment, however, because their relatively large sizes and signatures and limited maneuverability make them vulnerable to even modest air defenses.[4]

The Impact That the Proliferation of Unmanned Aerial Vehicles Has on the Combat Capabilities of and Interoperability with U.S. Allies and Partners

Most exports of U.S.-manufactured UAV systems have been to countries that are signatories to the MTCR agreement. The only nations to receive U.S. category I or near–category I UAV systems have been U.S. allies and partners. Some U.S. partners, such as Jordan, the UAE, and Saudi Arabia, were denied requests to purchase armed UAVs from the United States and have turned to China to purchase these systems (see, for example, Page and Sonne, 2017, and Scarborough, 2015). Some allies are also seeking foreign UAVs to avoid some of the operational conditions that are imposed when purchasing U.S. systems. In general, the proliferation of UAVs to U.S. allies and partners does provide them with additional capabilities.

---

[3] Our threat assessment is based mostly on U.S. capabilities and operational concepts; we assume that an opponent would use UAVs in the same general way as U.S. forces would (albeit with less effectiveness).

[4] This assessment does not consider more-advanced military UAVs (e.g., those with stealth capabilities) because they are not currently available for export and we do not have any information indicating that they will be exported in the future.

There are many advantages of allowing U.S. UAV manufacturers to sell UAVs to U.S. allies and partners. UAVs are valuable assets in achieving a variety of strategic, operational, and tactical objectives, including ISR missions and kinetic-strike operations. Close coordination and the ability to share UAV operations load is important for the United States in joint operations. This includes having the ability to control air vehicles and their subsystems, as well as the ingestion and integration of data. Tactical and operational interoperability is critical. UAVs have become the predominant tactical collection platform across all levels of command. This necessitates coordinating, sharing information to and from, and integrating UAVs into theater operations (Office of the Secretary of Defense, 2005).

If more allies and partners operate UAVs that are not interoperable with U.S. systems, combined warfighting will become less efficient. Overall, given proliferation and interoperability issues, we conclude that it is more beneficial to allow the sales of category I (and near–category I) UAVs to allies and partners. We determined that, although some risks are associated with selling U.S. UAVs to allies and partners (e.g., misuse and potential loss of technologies), there are significant advantages to the United States related to enhanced interoperability that these exports enable for potential future operations.

Although interoperability with partner UAVs is problematic (for potential cybersecurity, technical, and policy reasons) even when systems are U.S. manufactured, the associated challenges are easier to resolve when the UAV is built in the United States. High levels of interoperability are optimal, but work-arounds are currently in place to exercise command and control and to integrate data streams of partner UAVs. The United States is already doing this with U.S.-manufactured UAVs currently in use by the various service branches.

Missile Technology Control Regime Restrictions' Effect on Exports of Category I Unmanned Aerial Vehicles

We assess that the MTCR "strong presumption of denial" associated with category I (meaning that the transfer will occur on rare occasions and will be subject to additional agreements) has contributed to limiting the proliferation of these systems.[5] The United States has limited the export of category I UAVs to a small set of close allies, and most MTCR signatories have so far refrained from developing these large vehicles. However, the environment is rapidly changing as a capable global large-UAV industry is evolving. The United States is no longer the sole manufacturer of large UAVs. Non-MTCR nations, especially China, have been filling the void left by the United States and has been widely exporting near–category I systems, and China not only has been marketing a military category I UAV recently for export but is also building a manufacturing facility in Saudi Arabia to coproduce large UAVs, potentially including the category I CH-5 (with a payload capacity of 1,200 kg). Other nations, such as the UAE and, to a lesser extent, Germany and potentially Italy, are also starting to develop and export category I UAVs. The strict strong presumption of denial for category I UAVs could hinder the development of future large U.S. commercial UAVs and could be affecting U.S. companies' R&D investments and inhibiting those companies from pursuing some of these commercial opportunities.

---

[5] Additional agreements include assurances of responsibility from the recipient government for taking all steps necessary to ensure that the item is put only to its stated end use (MTCR, undated [c]).

Benefits and Risks of Continuing to Limit Exports

Although the MTCR restrictions influence export decisions, the regime is only one of many U.S. control mechanisms to which an export request is subjected. Therefore, like any other weapon system (and related technology), large UAVs will be subjected to export-control considerations regardless of what MTCR restrictions are in place. We evaluated the potential security, economic, and political consequences of maintaining the controls of category I UAV exports. In Table S.1, we summarize the assessments of the different factors discussed in this summary. The table shows that the MTCR's effects on security considerations are somewhat negative. Although this might appear counterintuitive because tight export controls are limiting U.S. systems from proliferating, U.S. allies and partners are being negatively affected while the threat to U.S. and allied troops from foreign-made UAVs, mostly from China, has increased. As expected, the economic impact on the United States is negative because fewer sales are occurring. If we consider the political results illustrated in the table, they would appear to be neutral. However, we should also consider more closely the relative risk and consequence of each factor (e.g., partner relationship versus misuse of UAVs), especially when assessing the political implications.

Looking across these assessments, we conclude that the MTCR's impact and effectiveness in controlling category I UAVs have eroded for several reasons:

1. A subset of category II systems, which we describe and label as near–category I, is widely proliferated. These systems have capabilities that are near those of category I systems but with smaller payloads (between 300 and 500 kg), so the strict category I restrictions do not apply.
2. Several nations are now developing and openly marketing category I systems for sale to both MTCR and non-MTCR nations.
3. Non-MTCR nations with the capability to manufacture category I systems (e.g., China) have also marketed these systems and, in one case, are building coproduction facilities to produce them.

Conclusions

The current pervasiveness of near–category I UAVs and the apparent future trend in the spread of category I systems lead us to conclude that the proliferation of large UAVs is ongoing and accelerating. Our analysis also suggests that the MTCR has had some effect on controlling the proliferation of category I UAVs. The United States has limited the export of these vehicles to a small set of allies, and most MTCR signatories have so far restrained from developing these large UAVs. None, to our knowledge, has yet exported them to non–MTCR signatories.

However, the landscape has changed dramatically in recent years. Non-MTCR nations, primarily China and, to a lesser extent, Israel, have exported large armed and unarmed near–category I UAVs, and China has recently openly marketed an armed category I system. Some non-MTCR nations have shown interest in building up their UAV capabilities and might either purchase these large UAVs or decide to coproduce them with China's help. Furthermore, some cracks appear to be forming within MTCR signatories. Germany is codeveloping a category I optionally piloted aircraft with Qatar, and Italy plans to export a potentially category I system to the UAE. We also conclude that the category I MTCR restrictions might be negatively affecting the capabilities of and interoperability with allies and partners.

Although the MTCR has been, and continues to be, an effective tool to control the spread of ballistic and cruise missiles, the availability of large UAVs from non-MTCR nations has significantly eroded the MTCR's efficacy to limit their proliferation. Furthermore, the MTCR's export restrictions associated with category I UAVs might be affecting current U.S. industry plans and investments in future large commercial UAVs.

Acknowledgments

This research was made possible through the support of our sponsor, Martin DeWing at the Joint Chiefs of Staff Office of Intelligence. We thank our action officers CDR Craig Fay and Lt Col Erynn Tait for the support, assistance, and guidance they provided throughout the term of the project.

We also want to thank COL Jeffrey Spears, for his feedback and help in providing us U.S. embassy contacts.

We thank John V. Parachini and Rich S. Girven at RAND for their support of this research and Lynn E. Davis and James Williams, our RAND colleagues, for their careful and thoughtful reviews of the report. This report has been much improved by their efforts. Finally, we are also grateful to Barbara Bicksler and Holly Johnson for their crucial assistance in the preparation of this report.

Abbreviations

C2	command and control
CASC	China Aerospace Science and Technology Corporation
COMINT	communications intelligence
ELINT	electronic intelligence
EO	electro-optical
FAA	Federal Aviation Administration
FMS	foreign military sale
GCS	ground control station
IAI	Israel Aerospace Industries
IR	infrared
ISR	intelligence, surveillance, and reconnaissance
LOS	line of sight
MALE	medium altitude, long endurance
MTCR	Missile Technology Control Regime
n/a	not applicable
NAS	National Airspace System
NATO	North Atlantic Treaty Organization
NDAA	National Defense Authorization Act
OPA	optionally piloted aircraft
R&D	research and development
RPA	remotely piloted aircraft
SAR	synthetic aperture radar
SATCOM	satellite communications

SEAD	suppression of enemy air defenses
SIGINT	signal intelligence
SIPRI	Stockholm International Peace Research Institute
STANAG	standardization agreement
UAE	United Arab Emirates
UAS	unmanned aircraft system
UAV	unmanned aerial vehicle
UK	United Kingdom
WMD	weapons of mass destruction

CHAPTER ONE

Introduction

The National Defense Authorization Act (NDAA) for Fiscal Year 2017, Section 1276, requires an independent assessment, directed by the Chairman of the Joint Chiefs of Staff, to report on the impact on U.S. national security interests of the proliferation of remotely piloted aircraft (RPA) that are assessed to be category I items (those subject to a strong presumption of denial for export) under the Missile Technology Control Regime (MTCR).[1] The NDAA requires this evaluation, in the form of a report, to be delivered to the congressional defense committees. The congressional language specifically requires the assessment to include evaluation in six areas:

> (1) A qualitative and quantitative assessment of the scope and scale of the proliferation of remotely piloted aircraft that are "Category I" items under the Missile Technology Control Regime. (2) An assessment of the threat posed to United States interests as a result of the proliferation of such aircraft to adversaries. (3) An assessment of the impact of the proliferation of such aircraft on the combat capabilities of and interoperability with allies and partners of the United States. (4) An analysis of the degree to which the United States has limited the proliferation of such aircraft as a result of the application of a "strong presumption of denial" for exports of such aircraft. (5) An assessment of the benefits and risks of continuing to limit exports of such aircraft. (6) Such other matters as the Chairman considers appropriate. (Pub. L. No. 114-328, 2016)

The RAND Corporation was tasked to help address the NDAA requirement by providing a direct response to each of these NDAA areas of evaluation. To help respond to these requests, we conducted literature reviews related to them, collected and analyzed publicly available and classified data and information, met with subject-matter experts to consider their expert feedback, and used these inputs to perform the requested assessments. In this report, we summarize and document the unclassified findings and assessment. Although we performed a comprehensive literature and data search and identified most of the major RPAs (more than 400), the universe of RPAs (to which we refer as unmanned aerial vehicles [UAVs] in the rest of the report) is extremely large, and data on some of them are not readily available.

Informed by our early findings, we decided to extend the types of UAVs to consider in this assessment beyond category I aircraft as requested by the NDAA. We included a subset of somewhat smaller UAVs that we call near–category I. Although these smaller systems do not fall under the MTCR category I purview, they can perform missions similar to those of the category I systems and pose similar threats to U.S. and allied forces.

---

[1]   Later in this chapter, we provide more detail on MTCR categories I and II.

The classified sources and results further support the assessments in this report. We discuss them in a separate annex with appropriate classification controls. Because they are not normally offered for export, we did not consider cruise missiles or any advanced UAVs—that is, military UAVs that incorporate stealth, self-protection, or other advanced capabilities (such as the X-47B).

To set the stage for the assessment and evaluation that follows, we started our review of RPA from a definitional perspective. In the rest of this chapter, we describe the major characteristics of RPAs, UAVs, and the MTCR.

Remotely Piloted Aircraft

This report applies to unmanned or uninhabited aircraft, whether remotely piloted or autonomous. These vehicles have various designations throughout the U.S. government and industry, including *RPA*, *UAVs*, and *unmanned aircraft*. The MTCR defines UAVs differently from current literature to include reconnaissance drones, target drones, and cruise missiles. To avoid misinterpretation, we use the most common term, *UAV*, rather than *RPA*, to refer to unmanned or uninhabited aircraft, which are the subject of this report.[2] The selection and use of this term have no effect on any of the assessments or findings in this report. The principal characteristics of a UAV or RPA are these (Davis et al., 2014):[3]

- It is a powered, aerial vehicle that does not carry a human on board.
- It uses aerodynamic forces to provide lift and maneuverability.
- It can fly autonomously with little to no direct human input, as a UAV, or piloted remotely with greater likelihood of human control, as an RPA. We note that, unlike cruise missiles, most currently deployed category I and II U.S. UAVs maintain communications with a command center where a human operator has ultimate control of the aircraft and maintains authority of weapon release.[4]
- It is recoverable—that is, it is designed and built to conduct multiple launch-and-recovery missions.
- It can carry a lethal or nonlethal payload.

Unmanned aircraft system (*UAS*) typically refers to a UAV and the supporting ground equipment needed to operate the aircraft.

Benefits and Limitations of Unmanned Aerial Vehicles

Current UAVs have numerous benefits that make them attractive to their users, with the most obvious being eliminated risk to the pilot. Other advantages include persistence (or endurance)—that is, being able to remain on station for long periods of time up to multiple

---

[2] Although *RPA* is a proper description of unmanned aircraft, it can have the connotation that a pilot is remotely flying the aircraft. Some UAVs can fly autonomously with minimal to no input from the controller.

[3] Technically, a UAV is similar to a cruise missile except that it is recoverable. Therefore, a UAV can be used as a cruise missile.

[4] Davis et al., 2014, does not specify the necessity of maintaining communication between a UAV and its command center.

days. For this reason, UAVs have been called the "unblinking eyes in the sky" ("Unblinking Eyes in the Sky," 2012). Furthermore, new UAVs with autonomous takeoff and landing are much easier to fly than manned aircraft are; they also have lower operational cost. UAVs offer design advantages because they do not have onboard human operations requirements (i.e., no space, physical limitations, or life support equipment for onboard pilots).

We note that current category I UAVs tend to be more vulnerable to air defenses and countermeasures than their manned counterparts are. We are not considering the more-advanced military UAVs (e.g., X-47B), which likely have much more–sophisticated capabilities than the typical UAVs that are exported have ("X-47B Unmanned Combat Air System," undated). Davis et al., 2014, contains a more in-depth discussion of the benefits and limitations of UAVs than we have here. We note that many of the characteristics of the UAVs considered in this report (e.g., long endurance, flight speed and altitude) make them ideal platforms for the fight against terrorism (e.g., intelligence, surveillance, and reconnaissance [ISR] missions), as well as low-intensity kinetic and nonkinetic missions.

The Missile Technology Control Regime

In 1987, the United States and several of its close allies formed the MTCR as a voluntary, informal political understanding that sought to limit the proliferation of missiles and missile technology (including cruise missiles) capable of delivering nuclear weapons.[5] The regime was later expanded to include systems able to deliver weapons of mass destruction (WMD) and has since grown to include 35 nations, shown in Table 1.1 (MTCR, undated [d]). Additionally, although China and Israel have not been invited to join the MTCR, Israel has made a statement that it will unilaterally adhere to the MTCR, and China has voluntarily pledged in the past to abide by the MTCR guidelines. However, it is clear, at least for China, that they do not intend to abide by the MTCR restrictions given their willingness to export large UAVs and their manufacturing facilities. Davis et al., 2014, presents an in-depth discussion of the MTCR and its relationship to UAVs. In summary, MTCR signatories implement license authorization requirements prior to exporting controlled items listed in the MTCR annex (see MTCR, undated [b]). The regime recognizes that some of the technologies it aims to control can have nonmilitary applications and therefore uses a case-by-case approach to determine whether an item in the MTCR list should be licensed for export. The control list defines two categories of items:

- **Category I** items include complete rockets and UAVs capable of delivering a payload of at least 500 kg to a range of at least 300 km, their major complete subsystems, and related software and technology (MTCR, undated [b]). Category I systems are subject to a "strong presumption of denial," meaning that they are exported only in rare situations.[6]

---

[5] Because the MTCR is an informal, voluntary agreement, there are no formal mechanisms for enforcement of the guidelines among signatories.

[6] It requires a binding government-to-government agreement ensuring that the item will be used for the stated purpose, will not be modified or replicated, and will not be retransferred, and the recipient government assumes responsibility for taking necessary steps to ensure that the item is put only to its stated end use (MTCR, undated [c]).

The MTCR, however, states that "the transfer of Category I production facilities will not be authorized."
- **Category II** items include other less sensitive and dual-use components, as well as complete systems capable of a range of at least 300 km, regardless of payload (see MTCR, undated [b]). The MTCR states that the export of category II systems "is subject to licensing requirements taking into consideration the non-proliferation factors specified in the MTCR guidelines" (MTCR, undated [b]). These factors include
  - concerns about proliferation of WMD
  - the significance of the transfer in terms of the potential development of delivery systems (other than manned aircraft) for WMD
  - the assessment of the end use of the transfers, including the relevant assurances from the recipient states
  - the risk of controlled items falling into the hands of terrorist groups and individuals.

The MTCR remains an important element of U.S. nonproliferation activities. However, over time, technology development in a range of related areas has placed strains on the MTCR.

Unmanned aircraft present a unique challenge to the MTCR. Their potential to be used as cruise missiles motivated their inclusion in the MTCR, while their operational uses and intents are very different from those of other MTCR delivery systems. UAVs' characteristics and capabilities are similar to those of manned aircraft; indeed, UAVs are replacing these aircraft for certain missions because of some of their unique advantages (e.g., long endurance makes them ideal for ISR for applications, and their relative ease and cost of operations make them desirable candidates for future commercial operations). There are motivations for maintaining UAVs as part of the MTCR (e.g., their potential to be used as primitive cruise missiles); however, there are also reasons for the MTCR to somehow ease control over UAVs (e.g., they

Table 1.1
Missile Technology Control Regime Signatory Nations

Year Joined	Signatory Nation
1987	Canada, France, Germany, Italy, Japan, UK, United States
1990	Australia, Belgium, Denmark, Luxembourg, Netherlands, Norway, Spain
1991	Austria, Finland, New Zealand, Sweden
1992	Greece, Ireland, Portugal, Switzerland
1993	Argentina, Hungary, Iceland
1995	Brazil, Russia, South Africa
1997	Turkey
1998	Czech Republic, Poland, Ukraine
2001	South Korea
2004	Bulgaria
2015	India

SOURCE: MTCR, undated (d).
NOTE: UK = United Kingdom.

are not intended for WMD delivery; they are of high value for many military operations, such as ISR; and their use in commercial applications (e.g., future cargo delivery) and as optionally piloted aircraft (OPA) has increased.

The United States recognized early on that including manned aircraft in the MTCR would be nearly impossible given the global market for these vehicles. Some also argued that "the speed and surprise that can be achieved with missiles were far greater than that achievable with manned aircraft and thus constituted a more potent form of nuclear delivery system" (see, for example, Bowen, 1997).

The 500-kg payload capacity was intended to represent the weight of a crude nuclear weapon, while the 300 km represented the minimum distance in a theater where a nuclear weapon might be used. The 500-kg payload weight limit is arbitrary in light of current and future missions of UAVs. A larger payload is useful from a capacity point of view (e.g., carry more conventional bombs for military missions and carry more cargo for a commercial application). However, a UAV with a payload capacity of less than 500 kg (say, 350 kg) can perform comparable ISR missions to those of its larger brethren. Similarly, a UAV with a smaller payload can conduct strike missions, albeit with a lighter weapon load. In other words, from a threat point of view, many category II systems (i.e., those with payloads less than 500 kg) are similar to those in the larger category I. However, for many future commercial applications, the capacity to carry large payloads is critical (e.g., cargo delivery, OPA, firefighting applications). Just as some automotive manufacturers and large retailers might be exploring autonomous cargo delivery, so are some UAV and aircraft manufacturers.

The study considered the MTCR's effects on the U.S. industrial base and commercial market. Although no commercial category I UAV is deployed today, UAV applications are rapidly growing, and companies are starting to consider and develop large UAVs for nonmilitary applications. Although that market is in its infancy, the MTCR can and does affect how U.S. companies invest for future commercial development (i.e., if they do not believe that they can export a commercial product, they may not invest in its development).

The response to the congressional questions has been framed with an eye toward these points of friction. Finding the proper balance between approving and rejecting exports will be essential for ensuring the continued relevance of the MTCR. Despite these issues, the MTCR continues to serve as an important nonproliferation tool.

An additional challenge to the MTCR in terms of limiting the use of unmanned aircraft as cruise missiles to potentially deliver WMD is the ability to convert manned into unmanned aircraft. Today's technologies allow a motivated nation to perform such conversions. The United States has been doing it since the end of World War II.[7] The United States and China have also converted older fighter aircraft, including early model F-16s and MiG-19s, into unmanned systems (Reed, 2013). Other examples of such conversions are also given in the restricted literature. All of these converted systems surpass the MTCR category I threshold. This is relevant because motivated nations would be able to perform this conversion in the future if their intentions were to develop a one-way delivery vehicle similar to a cruise missile.[8]

---

[7] The United States converted B-17 bombers into unmanned systems to collect air samples from nuclear tests and conduct raids on the enemy ("Drone B-17," undated).

[8] However, a converted manned aircraft might not be as effective as a cruise missile.

Additional Class of Unmanned Aerial Vehicle Defined for This Study

Informed by our early research findings, we were compelled to expand the types of military UAVs that should be considered in this study to include a subset of category II systems, which we designate as near–category I, that have payload capacities between 300 and 500 kg (and ranges greater than 300 km). We do so to better address the effects that large UAVs (i.e., category I and near–category I systems) have on military operations of the United States and its allies and provide a more comprehensive assessment of the MTCR. This additional category affects the first three areas requested in Section 1276 of the NDAA. The major reasons for considering this additional class of UAVs include the following:

- The 500-kg limit was selected in the early 1990s to address potential WMD that could have been developed at the time. Many current UAV payloads, both lethal and nonlethal, are significantly smaller than 500 kg, as discussed further in Chapter Two. Yet, despite their smaller payloads, the near–category I UAVs perform many of the missions that category I systems do. They are also significantly less expensive and pose nearly equivalent threats to U.S. national security (as discussed in Chapter Two).
- To avoid the category I strong presumption of denial, companies might have developed some long-range UAVs with payload capacities between 300 and 480 kg.[9] These UAVs, which we label near–category I, are subject to the looser category II review and are more widely proliferated.
- The near–category I UAVs are designed to have long endurance, 40 hours or more of continuous flight. Users might also have the option of trading some of the fuel used to loiter for additional payload capacity potentially exceeding 500 kg. Structural, volumetric, and load-distribution constraints might limit this trade.

This is not to say that the capacity to carry a larger payload is not important. Category I military UAVs offer an advantage over the near–category I systems in that they can carry larger sensors, providing better data and longer standoff distance, and carry more munitions. Several factors, including MTCR restrictions, unit cost, and mission needs, can drive the use of near–category I systems. Moreover, as sensors and even weapons become more miniaturized, the relative capabilities and, therefore, importance of near–category I platforms can be expected to become more substantial. As a result, we included these systems in our assessment to capture their potential threat, which is one of the key motives of this project.

Report Organization

We organized this report to address the first five requests specified in the 2017 NDAA. Chapter Two provides the results of a qualitative and quantitative assessment of the scope and scale of the proliferation of category I (and near–category I) UAVs, as defined by the MTCR. In Chapter Three, we summarize our assessment of the threat posed to U.S. interests as a result of the proliferation of these large UAVs to adversaries. In Chapter Four, we assess the impact of

---

[9]   *Strong presumption of denial* refers to the strict export control associated with category I systems, which are exported only in rare situations.

the proliferation of large UAVs on the combat capabilities of allies and partners of the United States and on the interoperability with allies and partners.

In Chapter Five, we summarize our analysis of the degree to which the United States has limited the proliferation of category I UAVs as a result of the application of MTCR restrictions. Chapter Six contains an assessment of the benefits and risks of continuing to limit exports of category I UAVs. In conducting this assessment, we considered impacts to national security, as well as to the U.S. industrial base and commercial interests. Finally, in Chapter Seven, we summarize our findings and provide an overall assessment based on the discussions provided in the report.

The report contains two appendixes. Appendix A includes a discussion of some of the technical details related to UAV category I determination, including trade-offs between range and payload. Appendix B contains tables of proliferation data relevant to our assessment.

CHAPTER TWO

An Assessment of the Scope and Scale of the Proliferation of Unmanned Aerial Vehicles

This chapter addresses the first request in Section 1276 of the 2017 NDAA: "a qualitative and quantitative assessment of the scope and scale of the proliferation of remotely piloted aircraft that are 'Category I' items under the Missile Technology Control Regime."

In this chapter, we provide data, where available, related to the level of proliferation of large UAVs in terms of the number of exports and the countries possessing these systems.

Introduction

For the purposes of this study, we assessed proliferation by measuring how many nations operate large UAVs, regardless of how many they have. Proliferation can occur through different means. We considered the three principal methods:

- importing a complete system (i.e., category I [and near–category I] UAVs)
- converting an available system (e.g., convert a manned system into a category I UAV or a category II UAV into a category I)
- acquiring (independently or with foreign assistance) the capability to develop and manufacture large UAVs indigenously.

The most expedient and common means to obtain a complete system is typically by importing it (assuming that the import process occurs within a reasonable time). Although a handful of nations have converted manned systems into UAVs, these systems maintain about the same capabilities as the original (i.e., approximate payload capacity, range, and endurance). Finally, several nations have begun to develop and manufacture large UAVs, some with foreign assistance and others indigenously. This capability could significantly increase proliferation if these nations do not abide by the MTCR and decide to export their indigenously produced large UAVs.

We reviewed and collected data related to the proliferation of both category I UAVs (those having ranges of greater than 300 km and payload capacities greater than 500 kg) and near–category I UAVs (those having ranges of greater than 300 km and payload capacities greater than 300 kg). Our review considered both MTCR signatory nations and nonsignatory nations. We deemed it important not only to understand what vehicles fall into the two MTCR categories but also to examine system capabilities, in terms of what types of missions they can perform or support.

To assess UAV proliferation, we performed a comprehensive literature search of UAV data sheets, publications, and official foreign-sale notifications. Further information from articles, U.S. government documents, industry feedback, and our quantitative analyses formed the basis of the qualitative assessment presented in this chapter.[1] We obtained category I export data from congressional notifications and the State Department, and we obtained most of the category II export data from the Stockholm International Peace Research Institute (SIPRI) database (SIPRI, undated [b]).

Missile Technology Control Regime Categories and Capabilities of Unmanned Aerial Vehicles

Our assessment of UAV proliferation considered MTCR category I and our definition of near–category I platforms. Per MTCR guidelines, the classification should be based on the maximum capability of the vehicle—in other words, the maximum payload the vehicle can deliver to 300 km (one way). We expect (as discussed later) that some UAVs have maximum payload capacities that, to avoid the MTCR export restriction, manufacturers might produce to exceed the advertised value.

Range Considerations

Range is one of two factors that determine whether the MTCR governs the exports of a given UAV platform as a category I technology. Understanding how UAV producers and export review committees define the range metric is important, as is assessing how the range restriction applies to U.S. and foreign UAV exports. The MTCR considers the range to be the maximum one-way distance that the UAV system can travel in stable flight over the earth's surface with a given payload weight. It is based on the design's maximum capabilities and is independent of external factors, such as operational restrictions or limitations imposed by telemetry or communication links.[2] Appendix A provides more details about the range metric and how it can vary based on payload weight and other assumptions.

The amount of fuel an ISR UAV requires is typically driven by its endurance, or how many hours it can be in the air. For this class of UAVs, the amount of fuel required for loitering is typically much greater than that needed to reach the one-way 300-km range that the MTCR specifies for category I vehicles. In other words, a portion of the fuel load could be traded for additional payload while maintaining the 300-km range capability.

Payload Weight Considerations

Payload is the second factor that determines whether the MTCR governs the exports of a given UAV platform as a category I item. An aircraft's payload capacity can be traded with fuel weight, subject to certain structural, launch and recovery, and vehicle control limitations (see Appendix A for a more detailed discussion), which can make determining exact capabilities based on claimed specifications difficult. Some foreign UAVs advertise a payload capacity between 400 and 480 kg and endurance of more than 20 hours. It is not clear to us whether

---

[1] The quantitative analysis involved collecting and categorizing data related to UAV exports and capabilities.

[2] However, current large UAV operations, unlike cruise missiles, require the aircraft to be in constant contact with its operator for positive control, so it cannot fly outside the range of the communication system.

some of these vehicles should be classified category II, as the manufacturer claims, or should instead be category I. Given the high-energy density of aviation fuel and the relatively high flight efficiency achievable by low-speed platforms, a relatively small amount of fuel would need to be traded for payload weight to qualify the system as category I.[3]

Mission Capability Considerations

Though not a factor in establishing MTCR restrictions, mission capabilities are a key consideration when considering utility and threat of UAVs. Category I platforms support ISR missions and can carry various sophisticated sensors, including electro-optical (EO)/infrared (IR), synthetic aperture radar (SAR), and laser designators. They can also be armed with various air-to-ground missiles and bombs. A significant number of category II systems, especially those we designate as near–category I, can also perform the same missions, albeit at a lower capacity (i.e., less munitions, smaller sensors).[4] Table 2.1 provides a comparison of basic mission capabilities of select category I and near–category I systems. The table shows that they have comparable capabilities and can perform similar military missions, including medium-altitude, high-endurance (MALE) ISR and strike missions. Thus, including near–category I systems in our assessment results in a more realistic and comprehensive assessment of mission capabilities.

Number of Missile Technology Control Regime Category I and II Unmanned Aerial Vehicle Systems

Figure 2.1 shows the number of known UAVs with advertised payload capacities greater than 100 kg and ranges of more than 300 km to address the question of the scale of proliferation of UAVs. Although the number of category I systems (14) is greater than the number of near–category I systems (ten), the known number of nations operating near–category I UAVs is approximately double those operating category Is. (See Appendix B for additional specifications for select systems.)

The MTCR does not directly differentiate between civilian and military UAVs because it considers only payload weight and range capabilities when categorizing systems. Although there are no current commercial category I UAVs, foreign companies are starting to develop large cargo UAVs. For example, a Chinese company is developing a cargo UAV able to deliver a payload greater than 1 ton to a range of 280 km (Glaser, 2017; Lin and Singer, 2017b). Interestingly, the reported range is slightly under the 300-km category I threshold. Industry representatives we contacted expect large commercial UAVs to play an important role in the future.

---

[3] The Heron TP-XP and Wing Loong II endurances are more than 30 hours, while the advertised payload capabilities are 450 kg and 480 kg, respectively. These systems would need to fly less than two hours to reach 300 km (i.e., need less than 10 percent of their fuel capacities). Therefore, up to 90 percent of the fuel could be traded for payload weight, subject to maximum takeoff weight and loading and flight control limitations (see Appendix A). Although fuel load figures for these systems are not available, we estimate them to be above 300 kg, which would easily allow these UAVs to trade more than 50 kg of fuel for payload, making them category I systems.

[4] After all, the 500-kg payload threshold specified for the category I system was established to address the lower weight of a crude nuclear weapon in the late 1980s, not UAV conventional payloads. Current technologies enable ISR sensors and air-to-ground missiles that weigh in the tens of kilograms.

Table 2.1
Comparison of Some Category I and Near–Category I Systems

Characteristic	CH		Heron TP	
	CH-4B (Near–Category I)	CH-5 (Category I)	Heron TP-XP (Near–Category I)	Heron TP (Category I)
Maximum payload, in kilograms	345	1,200	450	1,000
Weapon armament, type, and number	Four AR-1 antiarmor missiles (45-kg laser-guided bombs)	16 45-kg AR-1 antiarmor missiles, 20-kg AR-2 antiarmor missiles, and 50-kg FT-9 precision-guided munitions	Unknown	Spike and Tamuz (rumored)
Sensor list	EO/IR imager, SAR, electronic support subsystem, electronic countermeasures, self-protection jamming provisions, laser range-finding and illuminations, and communication relay equipment	EO/IR imager (with high-definition daylight charge-coupled device television camera, thermal imager, and laser range finder), electronic warfare equipment (SIGINT and jamming), SAR	EO/IR (optional), SAR, maritime patrol radar, ELINT, COMINT, electronic support measures	EO/IR (includes laser range-finding), SAR, maritime patrol radar, ELINT, COMINT, electronic support measures
Maximum altitude, in meters	8,000	7,000	13,700	13,700
Cruising speed, in kilometers per hour	150–180	180–220	Not provided by manufacturer	241
Endurance, in hours	14	39	30+	36
Range, in kilometers	1,600	2,000	>1,000	7,400

SOURCES: "ALIT/CASC CH-4 Series," 2017; Wong, 2016; Israel Aerospace Industries (IAI), undated (b).

NOTE: SIGINT = signals intelligence. ELINT = electronic intelligence. COMINT = communications intelligence.

Major Exporters of Unmanned Aerial Vehicles

UAV proliferation occurs principally through import, conversion, or indigenous production. The number of countries that have acquired UAVs (small and large) rose from 41 in 2004 to more than 76 in 2011 and is likely much larger today. In 2012, the U.S. Government Accountability Office estimated that more than 50 countries were developing more than 900 different UAVs (U.S. Government Accountability Office, 2012). The estimated global UAV market (military and commercial) is expected to grow from about $6 billion in 2015 to about $12 billion in 2025 (Stohl, 2015). A market research organization expects that commercial applications will be the fastest-growing sector, increasing from $512 million per year in 2015 to more than $6 billion in 2025 (Teal Group, 2017). These figures are estimated projections and include all UAV sizes; however, they point to a general growth in the future acquisition of UAVs and an expected increase in overall proliferation.

The demand for UAVs has increased rapidly in the past several years and is expected to continue to grow in the coming decade. According to recent work by the Stimson Center,

Figure 2.1
Number of Existing Unmanned Aerial Vehicle Systems

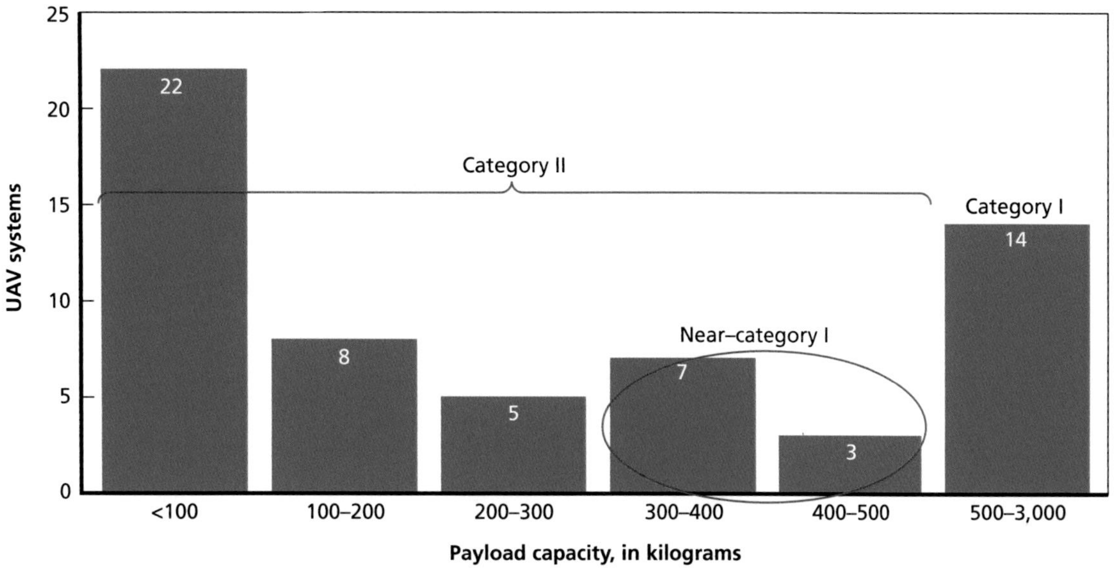

SOURCES: "All the World's Aircraft: Unmanned," undated; "Elbit Hermes 900 (Kochav) Medium Altitude, Long
Endurance (MALE) Unmanned Aerial Vehicle (UAV)," 2017; Lappin, 2017; Piaggio Aerospace, undated; "Heron TP
(Eitan) MALE UAV," undated; "Chang Hong-5 (CH-5) Combat and Reconnaissance Drone," undated; Aeronautics,
undated.
NOTE: The figure shows UAVs with ranges greater than 300 km at these payload weights.
RAND RR2369-2.1

industry experts calculated that exports from the United States alone totaled between $2 billion and $3 billion from 2012 to 2015 (Stohl, 2015).

A small handful of countries have been responsible for the majority of large UAV exports in recent years. The United States and Israel have been the leading UAV exporters for many years. Recently, China has become a large exporter, and other new exporters, such as the United Arab Emirates (UAE), could be on the horizon. Although the UAE is not considered a leading exporter, it is emerging as not only a potential major exporter of category I UAVs in the future but also a provider of manufacturing capabilities to nations interested in acquiring such capabilities. The next sections provide an overview of the export activities of these four countries, specifically as they relate to category I and near–category I UAV trades. Of these four countries, only the United States is a signatory of the MTCR, but Israel appears to abide by it.

United States

The United States has developed a variety of category I UAVs and has offered several for export—specifically, the Northrop Grumman RQ-4A Global Hawk and MQ-4 Triton and the General Atomics MQ-9 Reaper and Predator. As indicated in Table 2.2, these systems have been exported to North Atlantic Treaty Organization (NATO) members plus seven other countries. However, although many of these systems are category I, the United States has sold armed UAVs to only three countries: Italy, Spain, and the UK.

Recently, the United States has agreed to export category I systems to India. Specifically, in 2017, "the United States approved a $3 billion Foreign Military Sale (FMS) of 22 non-weaponized General Atomics SkyGuardian UAVs to the Indian Navy" ("US Approves Sale of

Table 2.2
Trades of Category I Unmanned Aerial Vehicles

Exporter	Importer	Number Ordered	System Name	Year of Delivery
United States	South Korea	4	RQ-4A Global Hawk	2017–2019
	France	16	MQ-9 Reaper	2013–2016
	Italy	4	MQ-9 Reaper	2010–2012
	Japan	3	RQ-4A Global Hawk	2019 (order cleared in 2015)
	NATO	5	RQ-4A Global Hawk	2016
	Spain	4	MQ-9 Reaper	Expected in 2019 (ordered in 2016)
	India	22	MQ-9 Guardian[a]	The deal is not complete but is likely to be approved.
	UK	11	MQ-9 Reaper	2007–2014
	UK	26	MQ-9 Predator B	Not yet delivered
	Australia	7	MQ-4C Triton	The order is likely complete; desired delivery is 2019.
Israel	Germany	5[b]	Heron TP/Eitan[c]	2018 (ordered in 2016)
UAE	Russia	At least 2	United 40[d]	2016
Germany	Qatar	17	Q01[e]	2017

SOURCES: Defense Security Cooperation Agency, undated; Haria, 2013; "PM Narendra Modi's Israel Visit," 2017; Reiner Stemme Utility Air Systems, undated (a); SIPRI, undated (a).

[a] The Guardian is the maritime variant of the MQ-9.

[b] SIPRI considers these data to be uncertain.

[c] The Heron TP, also known as the Eitan, is by IAI.

[d] United 40 is from ADCOM Systems.

[e] Reiner Stemme Utility Air Systems is the developer of the Q01.

22 SkyGuardians to India," 2017). Although the United States has allowed the sale to India, export controls restrict the United States from selling category I systems to several countries. Industry contractors mentioned that Saudi Arabia, the UAE, Jordan, Turkey, and Nigeria had all expressed interest in their systems but were turned down. Our records indicate that the United States has sold a near–category I system to only one nation, Switzerland (see Table 2.3).

Israel

From 1985 to 2014, Israel accounted for the majority (61 percent) of global UAV exports. Historically, Israel has not exported category I UAVs, but some of its systems come close to the category I threshold. Israel has exported a category I system only to Germany; however, it has exported near–category I systems, such as the Hermes 900, to more than five nations (see Table 2.2). In addition, "Kazakhstan is expected to shortly sign an agreement with Israel's Elbit Systems to undertake assembly of Skylark and Hermes UAVs" ("Kazakhstan to Start UAV Assembly in 2017," 2016). Also, Israel could soon export ten category I Heron TPs to India as part of a $400 million deal ("PM Narendra Modi's Israel Visit," 2017).

Table 2.3
Trades of Near–Category I Unmanned Aerial Vehicles

Exporter	Importer	Number Ordered	System	Year of Delivery
China	Iraq	4[a]	CH-4	2015
	Jordan	2[a]	CH-4	2016
	Algeria	Unknown	CH-4	Unknown
	Saudi Arabia	2[a]	CH-4	2015
	Saudi Arabia	Unknown	Wing Loong II	Unknown
	UAE	Unknown	CH-4, Wing Loong II	Unknown
	UAE	Unknown	Wing Loong II	Unknown
	Egypt	Unknown	CH-4	Unknown
	Egypt	Unknown	Wing Loong II	Unknown
	Nigeria	Unknown	Wing Loong II	Unknown
	Uzbekistan	Unknown	Wing Loong II	Unknown
	Kazakhstan	Unknown	Wing Loong II	Unknown
Israel	India	10	Heron TP-XP/Eitan[b]	Selected but not yet ordered
	Brazil	2[a]	Hermes 900[c]	2014
	Chile	3	Hermes 900	2013
	Colombia	1	Hermes 900	2014
	Switzerland	6	Hermes 900	Expected by 2020
	United Nations	3	Hermes 900	2016
Italy	UAE	8	P.1HH HammerHead[d]	Starting in 2018
United States	Switzerland	1	Centaur (OPA)[e]	2012

SOURCE: SIPRI, undated (a).
[a] SIPRI considers these data uncertain.
[b] The TP-XP is a special export version of the Heron TP.
[c] The Hermes 900 is from Elbit Systems.
[d] The P.1HH HammerHead comes from Piaggio Aerospace.
[e] Aurora Flight Sciences designed the Centaur.

China

China has rapidly filled the vacuum left by U.S. export rejections (Page and Sonne, 2017). China has exported near–category I UAVs (Wing Loong II and CH-4) to at least nine nations, as listed in Table 2.3. We argued earlier in this chapter that the Wing Loong II might qualify as a category I system. In addition, in July 2017, the China Aerospace Science and Technology Corporation (CASC) "announced that it was ready to mass-produce and offered its CH-5 drone for export" (Khan, 2017). The capabilities of this UAV place it above the category I threshold. Although no official sales have been reported, China is prepared to sell this category I UAV in the near future.

Moreover, China has also started to establish production lines with certain countries. Specifically, China has started to develop a production line in Saudi Arabia for up to 300 Wing Loong II and potentially the CH-5. It is not known whether Saudi Arabia will be able to export any UAVs it manufactures (Armstrong, 2017; Page and Sonne, 2017). This deal alone will provide Saudi Arabia with up to 300 category I and near–category I UAVs. Similarly, China has signed deals for potential production lines in Pakistan and Myanmar, although the status of that deal and which CH variant they will produce are unclear ("Saudi Arabia Imports UAV Production Line from China," 2017).

The United Arab Emirates

The UAE-based company ADCOM is a relative newcomer to the large-UAV export field. It has developed and sold the category I United 40 UAV, shown in Figure 2.2, to undisclosed customers, with potentially two of these systems going to Russia (Haria, 2013). ADCOM's client base appears to be growing rapidly, and the company claims that it will need to build a United 40 UAV every week to meet its current orders (Haria, 2013). It is also building a manufacturing facility in the UK and is targeting expansion in Saudi Arabia and India (Batey, 2015). We note that we could not confirm ADCOM's claims because information about the United 40 UAV is scarce.

Figure 2.2
The ADCOM United 40 Unmanned Aerial Vehicle

SOURCE: Photo by Vitaly V. Kuzmin via Wikimedia Commons (CC BY-SA 4.0).
RAND *RR2369-2.2*

Other Notable Development of Large Unmanned Aerial Vehicles

Germany is codeveloping the Q01 OPA with Qatar, shown in Figure 2.3. The unmanned capability of the Q01 exceeds the category I threshold (Reiner Stemme Utility Air Systems, undated [a]).

Italy has approved the export to the UAE (starting in 2018) of the P.1HH HammerHead UAV, shown in Figure 2.4. The HammerHead is capable of exceeding the category I threshold; however, it is not clear yet whether Italy will modify the design to constrain the payload to be below the 500-kg limit (Piaggio Aerospace, undated).

Other European efforts are underway to develop an advanced, large UAV; however, details are scarce, and there are no current indications that it will be exported outside the alliance (France, Germany, and Italy) developing it (Dassault Aviation, 2015).

Figure 2.3
The Q01 German–Qatar Optionally Piloted Aircraft, United Arab Emirates

SOURCE: Promotional image from Reiner Stemme Utility Air Systems.
RAND *RR2369-2.3*

Figure 2.4
P.1HH HammerHead Unmanned Aerial System, Italy

SOURCE: Promotional image from Piaggio Aerospace, undated.
RAND *RR2369-2.4*

Figure 2.5 illustrates the increasing exports of category I systems. However, we note that the figure does not capture the expected export of a large number of category I UAVs from both China and the UAE. Publicly available reporting indicates that both these nations have received orders for their category I UAVs; however, we do not have the data indicating how many category I UAVs have been ordered, nor do we know when they will be delivered. Figure 2.6 provides a geographical view of the proliferation of users of category I and near–category I systems.

Conclusions and Summary

Only a handful of countries currently operate category I UAVs, which could be due to MTCR restrictions, mission needs, or unit cost. However, manufacturers have been skirting the limits of the MTCR by exporting long-range vehicles with payloads between 300 and 500 kg. These near–category I vehicles are quickly proliferating, have mission capabilities similar to those of their category I counterparts, are significantly less expensive than the category I vehicles, and are not subject to the strong presumption of denial associated with category I systems. Figure 2.6 provides a global view of which nations operate large UAVs.

We expect, based on current findings, that the proliferation of large UAVs is accelerating, especially the near–category I systems and, to a somewhat lesser extent, the category I class. For example, there is potential for up to 300 new large UAVs in Saudi Arabia, as well as the unnamed recipients of the United 40 category I. China and the UAE are not only marketing

Figure 2.5
Growth in the Number of Exported Category I Unmanned Aerial Vehicles

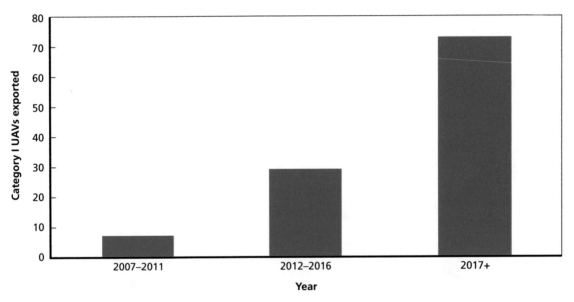

SOURCES: Defense Security Cooperation Agency, undated; Haria, 2013; "PM Narendra Modi's Israel Visit," 2017; Reiner Stemme Utility Air Systems, undated (a); SIPRI, undated (a).
NOTE: Numbers do not include any of the expected Chinese or UAE exports (except for two from the UAE to Russia).
RAND RR2369-2.5

Figure 2.6
Global Proliferation of Large Unmanned Aerial Vehicles

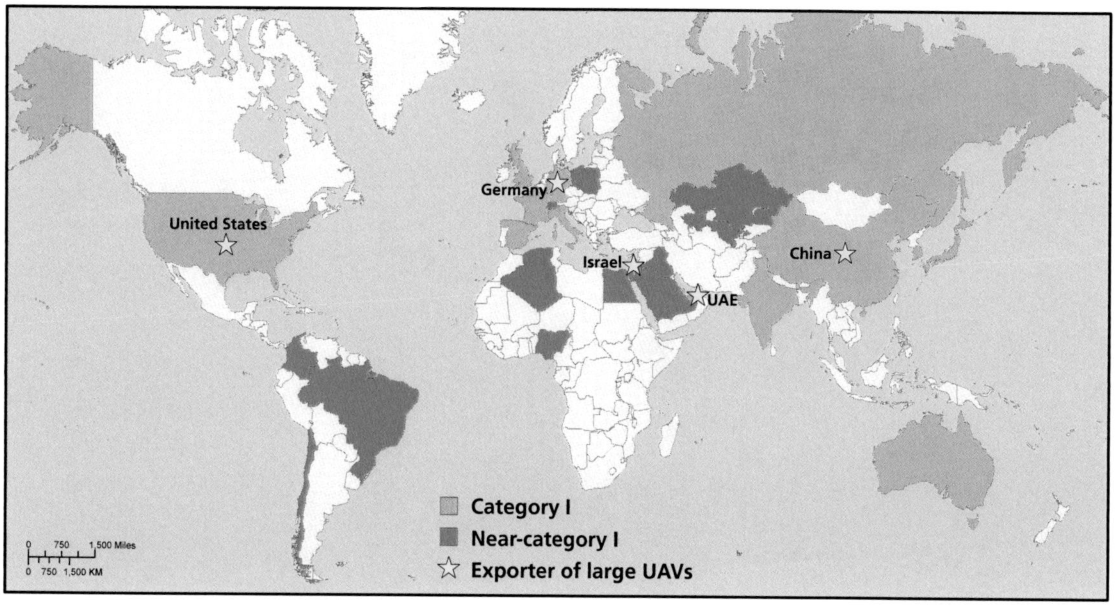

SOURCES: Defense Security Cooperation Agency, undated; Haria, 2013; "PM Narendra Modi's Israel Visit," 2017; Reiner Stemme Utility Air Systems, undated (a); SIPRI, undated (a).
RAND *RR2369-2.6*

large UAVs (including category I) but also offering to build factories for coproduction. These regional factories could further the proliferation of large UAVs to other nations.

CHAPTER THREE

The Threat That the Proliferation of Large Unmanned Aerial Vehicles Poses to U.S. Interests

This chapter addresses the second request in Section 1276 of the 2017 NDAA: "an assessment of the threat posed to United States interests as a result of the proliferation of such aircraft to adversaries." In this chapter, we consider and assess several representative missions in which UAVs could pose a threat to U.S. and allied forces.

The aircraft referred to in the NDAA are category I UAVs, as discussed in Chapter One. However, we also include near–category I systems (as described in Chapter One) because they do not face the stringent export restrictions that category I systems do, are more proliferated, and can, in many cases, pose similar threats to U.S. interests.

Regardless of U.S. or global policies regarding the export of UAVs, we see these systems posing an increasing potential threat to U.S. forces. Major potential adversaries—China, Russia, and Iran—recognize the utility of this capability and are producing many types of UAVs, including category I systems, for their armed forces (Lin and Singer, 2017a). Proliferation of these UAVs to less capable nations would, to some extent, complicate U.S. operations involving these adversaries. If export controls were successful in restricting some lesser adversaries from obtaining category I UAVs, they will likely be able to indigenously produce or acquire the near–category I systems (e.g., Saudi Arabia), as discussed in Chapter One.

Given the continuing evolution and proliferation of UAVs, it is likely that, in future conflicts, U.S. forces will have to cope with adversaries equipped with different types and sizes of UAVs, both armed and unarmed. Considerations include the following:

- Near-peer opponents will likely use a combination of category I–type UAVs, including advanced versions, such as the Chinese stealthy Lijian UAV, and smaller ones, depending on mission needs.
- Although some lesser opponents might not have category I UAVs, they will likely have near–category I systems, which they will attempt to use to fulfill many of the category I–type missions against U.S. forces.
- Lesser opponents might also develop UAVs by acquiring technologies through different means (available or illicit) and might also be able to convert manned into unmanned aircraft.

Unmanned Aerial Vehicles' Threats to U.S. and Allied Operations

Small and large UAVs pose various threats to U.S. interests abroad.[1] The potential threat that large UAVs pose to the U.S. homeland was beyond the scope of this study, and we do not address it in this report. UAVs, both large and small, are available in unarmed and armed configurations.[2] We based our threat assessment mostly on U.S. capabilities and operational concepts; we assume that an opponent would use UAVs in the same general way that U.S. forces do (albeit with less effectiveness).

Unarmed, sensor-centric, UAV missions include ISR, COMINT, ELINT, combat search and rescue, communication relay, maritime and border patrol, electronic warfare, and target acquisition and designation. Missions for armed UAVs encompass a wide variety of tasks currently undertaken by aircraft, missiles, or artillery, including counterterrorism strikes, suppression of enemy air defenses (SEAD), destruction of enemy air defenses, persistent hunter-killer or armed reconnaissance for time-sensitive targets, and antiship warfare (Office of the Secretary of Defense, 2005).

In the next sections, we explore in more detail five representative missions in which category I and near–category I UAVs can pose credible operational risks to U.S. and allied forces:

- operations against forward-deployed forces
- counter–special operations forces
- maritime domain awareness and sea control
- battlefield ISR and targeting
- support for SEAD and destruction of enemy air defenses.

These five cases are not exhaustive: UAVs can be effectively employed against U.S. and allied forces in other missions. The case studies are meant to highlight representative threats and relate them to what larger category I and near–category I platforms can accomplish, as well as to identify limitations in capabilities and employment of current systems. As the scenarios illustrate, both armed and unarmed UAVs pose a threat to U.S. interests.

We considered the potential of WMD delivery from UAVs as an additional mission area, but our analysis indicates that the threat from this type of UAV employment is relatively low. We conclude our threat assessment with a discussion on the suitability of using category I UAVs for WMD delivery.

Operations Against Forward-Deployed Forces

U.S. forward-deployed forces can be detected and surveilled from a variety of platforms, including MALE UAVs and high-altitude, long-endurance UAV systems posing the greatest threat. Medium-altitude systems are typically defined as having operating altitudes between 25,000 and 50,000 ft.; high-altitude systems operate above 50,000 ft.[3]

---

[1] Small UAVs can also present significant threats to U.S. interests; however, we do not address them in this report.

[2] Small UAVs can perform various missions, including ISR and kinetic strikes (e.g., China's CH-3, a small UAV with an 80-kg payload) can carry the AR-1 and AR-2 laser-guided missiles specifically designed for UAV applications (Fisher, 2016; Dominguez, 2017).

[3] *Long endurance* is typically defined as the ability to fly continuously for 12–24 hours or more. MALE systems include the Chinese Wing Loong II systems, the American MQ-9, and the Israeli Heron TP-XP. The RQ-4A Global Hawk is the only known operating high-altitude, long-endurance system and can fly at about 60,000 ft.

Both category I (e.g., the Chinese Wing Loong II systems, the American MQ-9, and the Israeli Heron TP) and near–category I (e.g., the Chinese CH-4 and the Israeli Heron TP-XP) systems have the ability to conduct persistent ISR of U.S. forces at standoff ranges. Operating from adjacent countries or in international airspace, an adversary would be able to conduct ISR operations against U.S. deployed forces to assess troop and equipment strengths, personnel and equipment readiness, and movements.[4] Although U.S. air surveillance systems would normally detect such operations (e.g., an adversary's preparation of the battlespace), the United States would likely be constrained in interdicting or interfering with the platforms prior to hostilities.

UAVs can be flown at lower altitudes when higher resolution of imagery data and full-motion video is desired, using EO/IR or SAR sensors, and to collect electronic and communication signals. Higher altitudes, on the other hand, allow aircraft to fly beyond the reach of some air defense systems, provide larger area coverage, and allow sensors to see deeper into a territory.[5] However, higher altitudes and longer surveillance ranges result in lower resolution and require larger antenna apertures to detect a given electronic signal.

During a conflict, weaponized UAVs would pose a threat to U.S. forces because they can carry their own weapons and laser designators to engage targets, including mobile ones. Some category I UAVs can be armed with ground attack weapons. The Chinese CH-5, for example, can reportedly carry up to 16 air-to-ground missiles, while the U.S. MQ-9 Reaper (shown in Figure 3.1) can be armed with a combination of missiles and lightweight guided bombs (Chen, 2017). The Israeli Heron TP is also capable of carrying several air-to-ground missiles. Several near–category I UAVs are weaponized, albeit with less firepower, such as the Chinese CH-4 (shown in Figure 3.2), American MQ-1, and Israeli Hermes 450. However, most UAVs, whether performing ISR or strike missions, are highly vulnerable to interdiction by even modest air defense systems (Davis et al., 2014, p. 13).

Counter–Special Operations Forces

UAVs equipped with EO/IR sensors and passive SIGINT systems can pose a threat to U.S. special operations forces, whose mission success normally depends on stealth and surprise. Adversary UAVs equipped with EO/IR sensors can provide persistent ISR and watch over likely lines of advance, landing zones, and choke points around target areas.

More-sophisticated UAVs with passive SIGINT payloads might be able to detect, triangulate, and exploit blue-force radio emissions. For some UAVs, the physical size of the radio-frequency receiver and onboard processor will likely limit their capabilities. And although one can argue that the threats that aerial passive detection poses can be mitigated by adopting sophisticated communication waveforms with low probability of intercept and low probability of detection, the existence of such capabilities could nonetheless impose operational constraints requiring U.S. forces to adapt.

Maritime Domain Awareness and Sea Control

UAVs can be deployed by adversaries to provide maritime domain awareness. Long-endurance and medium- and high-altitude UAVs equipped with appropriate sensors can provide persistent

[4]   An adversary country would be able to fly a UAV along the border of a country hosting U.S. forces without violating the host country's airspace.

[5]   As an example, a UAV flying at a 25,000-ft. altitude can cover a five-mile circular footprint. Doubling the altitude to 50,000 ft. allows it to cover a ten-mile circular footprint (assuming a 45-degree grazing angle).

Figure 3.1
Armed MQ-9 Reaper

SOURCE: U.S. Air Force photo/Lt. Col. Leslie Pratt.
RAND RR2369-3.1

surveillance of known shipping lanes or U.S. ships operating in international waters. Iran routinely uses UAVs to monitor movements of U.S. Navy forces in and around the Persian Gulf. UAVs can be equipped with sensors specifically designed to perform maritime surveillance.[6]

UAVs can also assist in exercising sea control. UAVs can provide long-term, persistent coverage of disputed areas and flashpoints. Their presence can be used to demonstrate sovereignty and resolve in peacetime while collecting intelligence in advance of potential conflicts. For example, China has reportedly deployed UAVs to monitor naval exercises in the East China Sea. It has also deployed the BZK-005 maritime surveillance UAV to coastal bases and to the disputed Paracel Islands (the BZK-005 does not meet the category I threshold) (Calderwood, 2016).

During a conflict, UAVs can provide an additional means to attack surface vessels, both by transmitting targeting information to other systems and by using onboard weapons. U.S., allied, and neutral shipping would be particularly vulnerable to detection and targeting by UAVs when operating in littoral and heavily navigated waters. Of interest is an *Aviation Week* article that reports that the United 40 UAV can purportedly be equipped with a torpedo for antiship warfare (Haria, 2013). However, during hostilities, UAVs supporting maritime operations are also vulnerable to air defense systems.

---

[6]   The MQ-4C Global Hawk is one example of a maritime surveillance UAV. See Northrop Grumman, undated (a).

Figure 3.2
Armed CH-4

SOURCE: Kelvin Wong/IHS.
RAND *RR2369-3.2*

Battlefield Intelligence, Surveillance, Reconnaissance, and Targeting Support for Long-Range Fires

As in maritime operations, UAVs can be employed for battlefield surveillance and targeting. These UAVs employ EO/IR sensors, as well as more-sophisticated ELINT and active ground moving-target indicator and SAR modes to locate and identify targets, use laser designators to facilitate engagement of targets, and then again use their sensors to assess effects of the engagements. Although smaller UAVs generally have less payload capacity and weapon potential than category I platforms, their ability to direct concentrated artillery or short-range rocket or missile fire makes them a serious threat during ground operations, as was demonstrated in eastern Ukraine (Rawnsley, 2015). The lower cost of smaller UAVs increases their utility in high-intensity ground combat, in which attrition is likely.

Support for Suppression of Enemy Air Defenses

Advanced adversaries can use UAVs to suppress or degrade air defenses. UAVs can be used to collect, target, and jam the electronic signals that are crucial to the operation of air defense systems. In Israel's Operation Peace for Galilee, UAVs collected critical electronic intelligence about surface-to-air missile emitters prior to hostilities and then used the data to stimulate air defenses in combat so that manned SEAD aircraft could attack them. Category I or near–category I UAVs would be effective in supporting this type of long-range or standoff SEAD support. Large UAVs might also deploy active decoys, such as the miniature air-launched decoy, at critical times with the intent of confusing and overwhelming an air defense system ("Raytheon to Equip GA-ASI's MQ-9 Reaper UAS with MALD," 2013).

Suitability of Unmanned Aerial Vehicles as Vehicles for Delivery of Weapons of Mass Destruction

Nefarious actors might consider UAVs as delivery vehicles for WMD. Given UAVs' capability to carry conventional missiles, nations with WMD could potentially arm UAVs with small nuclear or chemical missiles. However, there are disadvantages to doing so. Given the vulnerability of available UAVs to air defenses, arming them with WMD would subject their owners to significant risk of losing control of these weapons. States with WMD have deployed more-efficient and more-effective delivery means—missiles and manned aircraft—because they are significantly more survivable than most available UAVs are. Although some terrorist groups aspire to acquire WMD, very few have made progress, and they would be more likely to use simpler delivery methods. Moreover, there are significant limitations in employing UAVs to deliver WMD, as detailed in a previous RAND report (Davis et al., 2014, pp. 6–7).

As far as the use of large UAVs to deliver nuclear weapons, similar or potentially greater challenges apply. Nations pursuing nuclear weapons have more-effective and more-assured means of delivering these payloads via missiles and manned aircraft. A high-value payload, such as a nuclear weapon, would likely not be placed on a relatively low–survivable UAV platform.

Conclusion: Unmanned Aerial Vehicles' Threats to U.S. and Allied Operations

The proliferation of UAV systems represents an incremental but growing threat to U.S. and allied military operations. The primary threat from current category I UAV systems is their ability to conduct ISR operations against U.S. forces abroad prior to hostilities. Adversaries that would otherwise have difficulty detecting U.S. force deployments, monitoring U.S. operations, and maintaining targeting data on U.S. units can employ UAVs to maintain situational awareness of U.S. capabilities. Although near–category I (and some smaller) UAV systems can perform this function, the larger payload of the category I systems increases the weapon capacity of armed UAVs and enables larger, more-capable sensors that further increase the reach and endurance of UAVs conducting ISR. This capability is especially useful in monitoring ongoing force deployments in areas adjacent to coastlines or borders, where standoff surveillance is possible, and in the maritime domain.

During hostilities, battlefield surveillance, targeting, and lethal engagement are all potential threats from UAVs. Category I (and some category II) UAVs represent less of a threat in this environment, however, because their relative large sizes and signatures and their limited maneuverability make them vulnerable to even modest air defenses.[7] Smaller battlefield UAVs operating in larger numbers at low altitude constitute a more immediate threat to U.S. forces because U.S. close-in air defenses have degraded in recent decades (Freedberg, 2017).

---

[7]   This assessment did not consider more-advanced (e.g., stealthy) military UAVs.

CHAPTER FOUR

The Impact That the Proliferation of Unmanned Aerial Vehicles Has on the Combat Capabilities of and Interoperability with U.S. Allies and Partners

This chapter addresses the third request in Section 1276 of the 2017 NDAA: "an assessment of the impact of the proliferation of such aircraft on the combat capabilities of and interoperability with allies and partners of the United States."

In this chapter, we provide information regarding the current and future state of the UAV programs of U.S. allies and partners. In Chapter Two, we analyzed the proliferation of category I and near–category I UAV systems. This chapter seeks to explain why nations bought systems other than those sold by U.S. manufacturers and how these systems might affect allies' and partners' combat capabilities, as well as their interoperability with U.S. forces.

Ally and Partner Acquisition and Employment of Category I and Near–Category I Unmanned Aerial Vehicles

Most exports of U.S.-manufactured large-UAV systems have been to countries that are signatories to the MTCR agreement. As discussed in Chapter Two, the only nations to receive U.S. category I or near–category I UAV systems have been U.S. allies and partners. Several of these countries, however, have also acquired foreign-made UAV systems, as shown in Table 4.1.

Some U.S. partners, such as Jordan, UAE, and Saudi Arabia, unable to purchase armed UAVs from the United States have turned to China to purchase these systems (see, for example, Page and Sonne, 2017, and Scarborough, 2015). Chinese systems, such as the CH-4B and Wing Loong I, both of which are part of Jordan's and Saudi Arabia's UAV arsenals, provide capabilities comparable to those of the category II U.S. MQ-1 Predator, including multi-intelligence sensor suites and armament (Rawnsley, 2016; "ALIT/CASC CH-4 Series," 2017; "AVIC Wing Loong Series," 2017).

Some allies are also seeking foreign UAVs to avoid some of the operational conditions that are imposed when purchasing U.S. systems. France, for example, is required to obtain the consent of the United States to deploy its armed MQ-9 Reaper UAV systems and has expressed interest in developing jointly with Germany, Italy, and Spain by 2025 an indigenous UAV program to reduce reliance on U.S. and Israeli systems (Sayler et al., 2016; Cohen, 2017; Emmott, 2015).[1] Current plans for the program include civil and military applications, such as persistent surveillance and a strike capability (Pomerleau, 2015).

---

[1] The restrictions imposed on France (i.e., obtaining U.S. permission before deploying U.S.-made UAVs) are among the reasons France and other nations are seeking alternative sources for UAVs, including indigenous sources.

Table 4.1
U.S. Ally and Partner Nations with U.S. and Non-U.S. Large Unmanned Aerial Vehicle Systems

U.S. Ally or Partner Nation	U.S. UAV System	Non-U.S. UAV System
Republic of Korea	RQ-4 (ordered)	
France	MQ-9	
Spain	MQ-9	Flyox I[a]
Germany		Q01, Heron TP (ordered)
Italy	MQ-9	P.1HH HammerHead
Japan	RQ-4 (ordered)	
U.K.	MQ-9	
NATO	RQ-4	
India	MQ-9 Guardian (ordered)	Heron TP/XP (ordered), Rustom[b]
Australia	MQ-4C (ordered)	
Israel		Heron TP, Super Heron[c], Hermes 900, Dominator[c]
Poland		ILX-27
UAE		Wing Loong I, P.1HH HammerHead (ordered), CH-4, United 40, Hazim 15 A/B[e], Yabhon Flash-20, Yabhon Smart Eye
Switzerland	Centaur	Hermes 900 (ordered)
Iraq		CH-4
Jordan		CH-4

SOURCE: SIPRI Arms Transfers Database, undated (a) (data queried July 17, 2017).
[a] Singular Aircraft makes the Flyox I.
[b] The Rustom is a product of the Defence Research and Development Organisation.
[c] The Super Heron is by IAI.
[d] Aeronautics Defense makes the Dominator.
[e] ADCOM Systems makes the Hazim 15.

Allies' and Partners' Capabilities Enabled by Unmanned Aerial Vehicles

In general, U.S. ally and partner nations employ UAV systems for persistent ISR over land and coastal areas in countries without effective antiaircraft capability. They also use kinetic air-to-ground missions for national defense and global counterterrorism operations. The UK, for example, is the oldest user of armed U.S. UAV systems and has flown thousands of tactical reconnaissance and combat missions in Iraq and Afghanistan, including close air support, video support for surface operations, direct-fire support, interdiction, strategic attack, battle damage assessment, and support for combat search-and-rescue operations ("United Kingdom: Air Force," 2017). National airspace restrictions do not allow current UAV systems to operate in the UK; however, the United States is set to deliver the certifiable Predator B UAV, which meets UAV-airworthiness NATO Standardization Agreement (STANAG) 4671 to be able to fly in the UK national airspace (Apthorp, 2017). Australia, Italy, and NATO have been employing imported UAVs in Afghanistan. Australia has also flown them on a trial basis with its Border Protection Command (see, for example, "IAI Heron," 2017).

India has incorporated UAVs in all three of its military branches and has employed UAVs in regional conflicts—for example, by conducting reconnaissance in contested airspace along the Line of Control in the Jammu and Kashmir region along the India–China border ("India: Air Force," 2017; "India Deploys US-Made Surveillance Drones Along LoC," 2017). Iraq has also been using its Chinese-made UAVs to execute counterterrorism reconnaissance and strike operations against Islamic State of Iraq and Syria strongholds in Iraq, and the Jordanian Air Force acquired armed Chinese UAVs to use in its intensified fight against Islamic State of Iraq and Syria militants in Syria ("ALIT/CASC CH-4 Series," 2017).

Other countries, such as France and the UAE, employ their UAVs to support limited coalition missions in Africa and the Middle East. France's U.S. UAVs have been conducting ISR and target acquisition and tracking in the United Nations Multidimensional Integrated Stabilization Mission in Mali and kinetic strikes against militants in the Sahel–Sahara region for Operation Barkhane ("France: Air Force," 2017). Furthermore, the UAE has stationed its Wing Loong UAVs in Saudi Arabia to support the Saudi-led coalition in Yemen, and in Libya to support the Libyan National Army in its fight against Islamist militants ("United Arab Emirates: Air Force," 2017).

In the foreseeable future, U.S. allies and partners will likely strive to expand and improve the capability of their UAV fleets. Proliferation and ongoing development trends indicate increased interest in more-survivable, multimission, ISR, and combat platforms, capable of collecting and intercepting communications and electronic signals, executing electronic attacks, and employing electronic countermeasures, as well as precision bombing against ground and maritime targets and air-to-air missile engagements. Switzerland and several European Union nations, for example, have collaborated on an armed UAV technology demonstrator incorporating stealth (see, for example, Kreps, 2014).

An obvious advantage to allies and partners acquiring capable large UAVs is their ability to use them in combined operations and alleviate U.S. requirements for UAV operations.

Unmanned Aerial Vehicles' Interoperability Among Allies and Partners

It is important, if not critical, for U.S. and partners' UAVs to be able to coordinate and operate in common areas. This coordination requires some level of interoperability between U.S.- and partner-owned UAVs. Maintaining situational awareness, enabling rapid deconfliction, allowing desired sharing of command and control (C2) functions, and ensuring effective distribution of sensor and payload data between partners are all important goals that advance U.S. national security interests (see Stohl, 2015, and Schulberg, 2014).

There are both technical and organizational challenges to enabling interoperability of UAV missions between the United States and its partners (regardless of the size of the UAV). For example, even if a foreign partner is operating a U.S.-manufactured UAV system, that UAV might not be allowed to interface directly with U.S. ground control stations (GCSs) or ISR distribution centers because of U.S. policies (independent of the MTCR).

There are advantages and disadvantages to interoperability that could follow from partners acquiring U.S. or non-U.S. UAVs. There is the potential that interoperability benefits could result from relaxing UAV export controls within the MTCR and allowing more allies and partners of the United States to import large and technically advanced U.S.-made UAV systems. Yet other challenges, discussed in the next section, could reduce these benefits.

Levels of Interoperability

NATO STANAG 4586, which defines UAV interfaces (Marques, 2015), describes five levels of UAV interoperability, as listed in Table 4.2. The lower levels of interoperability cover dissemination of sensor and payload data between partnered operators, while the higher levels of interoperability relate to the sharing of C2 capabilities between operators. The technological barriers to achieving any level of interoperability are significantly reduced or completely removed if an ally acquires U.S. UAV systems; however, different standards for signal encryption or software requirements can remain (Mayer, 2017).

The State of U.S. Systems' Interoperability with Allied, Partner, and Other Systems

The current level of interoperability between UAV systems operated by the United States and those operated by its partners is low. No direct distribution of payload or sensor data is conducted between foreign-operated UAVs and U.S. control or analysis stations. Although foreign analysts can be physically located in U.S. analysis stations and, in some cases, distribute ISR to their respective coalition forces (see, for example, "Distributed Common Ground System [DCGS]," 2017), this is the current maximum extent of this type of interoperability between the United States and its partners.

The low level of interoperability might be due to technical limitations or policy and protocol that govern the interactions between the United States and its partners. If the current limitations are technological, allowing more exports of U.S.-manufactured UAV systems could quickly increase the level of U.S. and ally interoperability. If policy choices and organizational realities also limit the interactions between U.S. and foreign systems, allowing U.S. allies to use U.S.-manufactured UAVs more extensively would not necessarily increase interoperability without modifying policies to allow closer interactions.

Allied and Partner Systems

It is particularly important to consider interoperability between U.S. systems and UAVs exported by U.S. allies and partners. Currently, even foreign-operated UAVs *that were built by U.S. manufacturers* do not directly interface with U.S. control stations and Distributed Common Ground System centers. There are organizational and security challenges to allowing foreign-operated UAVs to interface with U.S. military infrastructure (e.g., potential cyber-related threats) even with very few or no technical restrictions. Foreign allies' and partners' use of U.S.-made UAVs should facilitate the United States and its partners leveraging the data because U.S. personnel understand the capabilities of the platform and its sensors.

Table 4.2
Levels of Interoperability of Unmanned Aerial Vehicles

Level	Description
1	Indirect receipt or transmission of sensor product and associated metadata
2	Direct receipt of sensor product data and associated metadata from the UAV
3	Control and monitoring of the UAV payload, unless specified as monitor only
4	Control and monitoring of the UAV, unless specified as monitor only, less launch and recovery
5	Control and monitoring of UAV launch and recovery, unless specified as monitor only

SOURCE: Marques, 2015.

Several standards, primarily from NATO, promote system architectures and practices that enable interoperability between different allied UAV systems. The two most important are the previously discussed NATO STANAG 4586, which provides standards for UAV control systems, and NATO STANAG 7085, which governs the data link for certain payloads (NATO, 2004). Although the use of U.S.-manufactured products identical to those that the U.S. armed forces use could lead to the highest-level and most-natural interoperability between platforms, leveraging standards such as these creates other opportunities to enable interoperability between U.S. systems and UAVs that are exported by U.S. allies.

For example, General Atomics and Lockheed Martin in the United States, as well as IAI and Elbit in Israel, have publicly announced their compliance with NATO STANAG 4586 (General Atomics Aeronautical, undated [a]; Lockheed Martin, undated; IAI, undated [a]; Elbit Systems, undated [a]). Thales, a French company, provided a data link that conforms to NATO STANAG 7085—a standard that is restricted to NATO members—to Israel for flight-testing on an IAI Heron UAV (Carey, 2015). This demonstrates that Israel, a U.S. ally and major UAV exporter, can create UAVs that comply with NATO interoperability technical standards. Meeting these standards, however, does not necessarily ensure interoperability. Standards for C2 communication and data links are mature, but connecting the various sources of information remains a largely unresolved challenge (Mayer, 2017).

Although Israeli UAV platforms have demonstrated some compliance with interoperability standards, allies and partners using Israeli- or U.S.-manufactured UAVs cannot currently interface directly with U.S. C2 and ISR distribution centers for UAVs. Increasing interoperability between U.S. systems and UAVs flown by allies and partners might have a technical advantage in terms of more-frequent use of U.S.-exported UAVs, but changes to policy and organizational structure are also required.

Other Systems

Similar challenges exist when considering interoperability between U.S. UAVs and systems that were designed and built by non–U.S. allies, such as China. There is significantly less public information surrounding these systems, but there is no indication that Chinese exported systems are intended to comply with international standards, such as the NATO STANAGs. Policies governing the interfaces of systems designed and built by non–U.S. allies must also consider additional risks in communication security resulting from relatively poor understanding of the systems, as well as potential exploitation by opponents. There are likely much more–substantial challenges to enabling interoperability between U.S. operational systems and UAVs manufactured by non–U.S. allies, including technological barriers for interoperability and potential cyberthreats introduced through these systems.

Currently, there is relatively little incentive for interoperability between U.S. systems and UAVs that were built by non–U.S. allies. However, U.S. partners employ Chinese UAVs in the same region in which the United States operates, which might change the incentives to increase interoperability in the future. As an example, the Iraqi military has reportedly used a combat-capable Chinese UAV during counterterrorism operations (Rawnsley, 2016). If Iraq increases its use of foreign-made UAV systems and the United States continues to partner closely with Iraqi forces in the region, achieving some level of interoperability between Iraqi and U.S. UAV systems might be needed; however, there would likely be some significant challenges to overcome before close interoperability with Chinese systems can be accomplished. Nations that have purchased foreign-made UAVs currently operate in the same region as the

United States does in conflicts in Yemen, Saudi Arabia (Blanchard, 2017), Libya, and other areas, which creates risks to having multiple countries operating in the same theater with limited to no interoperability capabilities. Some of these issues could be addressed by allowing select partners to purchase U.S. UAVs.

If China becomes a principal supplier of UAV technology to U.S. allies, the incentive to incorporate these systems into the U.S. operational environment would undoubtedly increase. There is an increased risk of greater Chinese UAV proliferation to allies if the United States continues to limit the export of large UAVs. In the foreseeable future, however, there will likely be significant technical, policy, and security (e.g., cyber-related) challenges to incorporating Chinese and other nonally UAVs into a U.S.-led alliance.

The Outlook for Future Interoperability Between U.S.-Supplied Systems and Ally and Partner Systems

As UAVs proliferate, there will likely be more incentives to increase interoperability between foreign-operated UAVs and U.S. military systems and organizations. The export policy regarding U.S. UAVs will affect what systems U.S. allies will be using and will likely influence the future level of interoperability.

In the foreseeable future, U.S. coalition allies that operate in tandem with the U.S. military will overwhelmingly be using UAVs built by the United States and its allies (such as the U.S.-built MQ-9 and the Israeli-built IAI Heron TP-XP) rather than those exported by China (such as the CH-5). An important question regarding future interoperability is whether the increased export and use of U.S. UAV systems will result in more-effective interoperability than continued reliance on Israeli or other allied UAV systems. It is, however, possible for ally-built UAV systems to meet many of the technical interface requirements (e.g., STANAG) to enable better interoperability with U.S.-built systems.

Probably, operational effects of increasing the use of U.S.-manufactured UAVs could be realized in the future. Being able to leverage both U.S. and allied personnel, logistics, and materiel in support of U.S.-made UAVs could further enhance the effectiveness of U.S. UAV systems. This will be achieved much more easily if U.S. allies and partners are using UAVs manufactured in the United States. Depending on the level of integration between U.S. and coalition forces, greater ally and partner use of U.S.-manufactured systems could create a significant advantage to the logistical support surrounding the maintenance and operation of coalition UAV systems. On the other hand, if U.S. partners increase their use of UAVs made by other nations, the logistical complexity to support the diverse array of UAV platforms in an allied operation will increase.

As discussed earlier, there are both technical and protocol challenges to interoperability between U.S.- and partner-operated systems. Currently, policy and organizational structures limit the direct interface between foreign-operated UAVs and U.S. systems. This is the case even when a U.S. company manufactured the foreign-operated UAV or the system is otherwise technically capable of interfacing with U.S. systems. If U.S. UAV exports are intended to increase the level of interoperability, it is essential to consider whether organizational and policy challenges override perceived technical benefits. In the future, if organizational schemes are put in place to enable UAV interoperability, increasing the export of large U.S. UAVs would make it easier to achieve higher levels of interoperability.

With no direct C2 or sensor data-link communications between U.S.- and foreign-operated systems, the current level of interoperability between U.S. and partnered UAV systems is low. Although increased allied export and use of U.S.-manufactured UAVs might increase interoperability by reducing technological barriers, there are other options to overcoming this challenge (e.g., compliance with NATO STANAGs), and potential policy or organizational challenges are likely more limiting than technical challenges. Achieving higher levels of interoperability would require concurrent modifications to U.S. C2 and ISR organizations' data ingestion and potential connectivity policies. Table 4.3 shows the most-significant potential advantages and challenges to increased use of U.S. UAVs, as well as the risks of allowing non-U.S. UAVs to be increasingly used.

Conclusions and Implications

There are many advantages of allowing U.S. UAV manufacturers to sell UAVs to U.S. allies and partners. UAVs are valuable assets in achieving a variety of strategic, operational, and tactical objectives, including ISR missions and kinetic-strike operations. Allies and partners would therefore benefit from acquiring UAVs, regardless of whether they are made in the United States. However, we assess that it would be more beneficial for the acquired UAVs to be U.S. made because they would help combined operations by facilitating logistical support and data distribution. Close coordination and the ability to share UAV operations load are important for the United States to leverage UAVs in joint operations. This includes having the ability to control air vehicles and their subsystems, as well as to ingest and integrate data. Tactical and operational interoperability is critical. UAVs have become the predominant tactical collection platform across all levels of command. This necessitates coordinating, sharing information to and from, and integrating UAVs into theater operations (Office of the Secretary of Defense, 2005).

If more allies and partners begin to operate UAVs that are not interoperable with U.S. systems, joint warfighting will likely become less effective. Overall, given proliferation and

Table 4.3
Advantages, Challenges, and Risks of Interoperability of Unmanned Aerial Vehicles

Increased Partner Use of U.S. UAVs		Challenges in Increased Use of	
Advantages	Challenges	Partner-Made UAVs	Non–Partner-Made UAVs
Significantly reduced technological barriers to direct ISR distribution	Additional policy and organizational challenges to enable interconnectivity and high levels of interoperability	Dependence on international standards (e.g., NATO STANAGs) to enable technological interoperability	Significant policy, organizational, and security challenges, including cyber-related risks, to interconnectivity
Significantly reduced technological barriers to C2 sharing, if desired		Additional policy and organizational challenges to interconnectivity	Risk that non–partner-made systems become commonly used by coalition partners
Leverage allied support (e.g., logistical, materiel, personnel) for U.S. UAV systems		Complexity of logistical support for diverse UAV platforms	Necessity of greater integration of non–partner-made UAVs introducing new security and organizational risks
Increased understanding of UAV platform and ISR capabilities among partners			Complex logistical support for diverse UAV platforms

interoperability issues, we conclude that it is more beneficial to allow than prevent the sales of category I UAVs to allies and partners. We determined that, although some risks are associated with selling allies and partners U.S. UAVs and the technology inherent in them, there are significant advantages related to the enhanced interoperability that these exports enable, as summarized in the first column of Table 4.3.

Although interoperability with partner UAVs is problematic even when systems are U.S. manufactured, the associated challenges are easier to resolve when the UAV is U.S. built. High levels of interoperability are optimal, but work-arounds are currently in place to exercise C2 and integrate data streams of partner UAVs when interoperability is problematic. The U.S. military is already doing this with U.S.-manufactured UAVs currently in use by the different U.S. service branches.

CHAPTER FIVE

The Effect That the Strong Presumption of Denial Has on Exports of Unmanned Aerial Vehicles

This chapter addresses the fourth request in Section 1276 of the 2017 NDAA: "an analysis of the degree to which the United States has limited the proliferation of such aircraft as a result of the application of a 'strong presumption of denial' for exports of such aircraft."

In this chapter, we evaluate how much of an impact the MTCR has had on the export of large UAVs.

The Effect of Missile Technology Control Regime Category I Restrictions on U.S. Exports of Unmanned Aerial Vehicles

The MTCR is one of several control mechanisms to which U.S. export requests of certain military, or dual-use, equipment are subjected. The MTCR is invoked when the request involves UAV and cruise missile– or ballistic missile–related technologies and systems. Although the MTCR has been, and continues to be, an influential export-control regime, some export denials are caused by factors independent of the MTCR, such as UAS export policy.[1]

The MTCR establishes a strong presumption to deny transfers of category I UAVs "to any destination beyond the Government's jurisdiction or control."[2] Such transfers "will be authorized only on rare occasions" and—among other requirements—with the supplier and not just the recipient taking "all necessary steps to ensure that the item is put only to its stated end use."[3]

---

[1] Chapter Two of this report notes the preponderance of foreign UAVs with range or payload limits just below the category I restrictions.

[2] See MTCR, undated (a), for the guidelines (rules) and annex (items to which the rules apply). The MTCR annex defines category I UAVs as "complete unmanned aerial vehicle systems (including cruise missiles, target drones and reconnaissance drones) capable of delivering at least a 500 kg 'payload' to a 'range' of at least 300 km."

[3] The MTCR guidelines set out these rules for category I transfers:

These Guidelines, including the attached Annex, form the basis for controlling transfers to any destination beyond the Government's jurisdiction or control of all delivery systems (other than manned aircraft) capable of delivering weapons of mass destruction, and of equipment and technology relevant to missiles whose performance in terms of payload and range exceeds stated parameters. . . . Particular restraint will be exercised in the consideration of Category I transfers regardless of their purpose, and there will be a strong presumption to deny such transfers. . . . Until further notice, the transfer of Category I production facilities will not be authorized. The transfer of other Category I items will be authorized only on rare occasions and where the Government (A) obtains binding government-to-government undertakings embodying the assurances from the recipient government called for in paragraph 5 of these Guidelines and (B) assumes responsibility for taking all steps necessary to ensure that the item is put only to its stated end-use.

Given that the United States retains the sovereign ability to export UAVs under the MTCR, notwithstanding the strong presumption of denial, the United States has approved a significant number of category I UAV exports—86 since 2006—all to allies that are also signatories of the MTCR, as shown in Table 5.1. The table also shows that 22 more are expected to be approved (given a favorable advisory opinion by the Department of State) for export to India. However, approvals do not imply actual sales. As Table 5.1 indicates, Germany and the Netherlands were approved for MQ-9 exports; however, they decided to not pursue these purchases. Additionally, some sales turned out to be smaller than the full amounts approved.[4]

To fully assess the MTCR's impact on U.S. exports, we also need to consider export denials. All category I UAVs are required to go through the FMS process, which is typically initiated by a letter of request or a pre–letter of request.[5] Denials are often handled by not

Table 5.1
U.S. Approvals of Exports of Category I Unmanned Aerial Vehicles

Year	Importer	Number	System	Approved Value, in Millions of Dollars
Congress notified				
2006	UK	10	MQ-9	
2008	Germany	5	MQ-9	0[a]
2009	Italy	2	MQ-9	63
2013	Republic of Korea	4	RQ-4	1,200
2013	France	16	MQ-9	1,500
2015	Netherlands	4	MQ-9	0[a]
2015	Spain	4	MQ-9	243
2015	Italy	2	MQ-9 weaponization	130
2015	Japan	3	RQ-4	1,200
2016	UK	26	MQ-9	1,000
Department of State approved but Congress not notified				
2012	NATO	5	RQ-4	1,700
Favorable advisory opinion				
2017	India	22	MQ-9B Guardian	2,000?[b]
In process				
	Australia	7	MQ-4C Triton	

SOURCE: Defense Security Cooperation Agency, undated.
[a] Germany and the Netherlands canceled their orders after approval.
[b] The approved value for India for 2017 is unknown until the State Department issues its advisory opinion.

---

[4] A U.S. manufacturer noted that the actual sales can be 25 to 50 percent less than the approved exported values.

[5] The UAS export policy requires that U.S. UAV sales go through FMS (U.S. Department of State, 2015). FMS requires that the importing nation work with the U.S. Department of Defense to acquire a weapon system. An importing nation is

responding to the request. We could not determine whether official records existed indicating which or how many requests for exports were denied or not addressed (effectively rejecting the request). However, industry representatives indicated that a large number of category I requests go unanswered and that many requests are never initiated because the interested parties believe that their requests will be rejected.

The Effect That Missile Technology Control Regime Category I Restrictions Have on International Exports of Unmanned Aerial Vehicles

We assess, based on the findings summarized in Chapter Two, that the MTCR controls have contributed to the current limited proliferation of category I systems. The United States, for its part, has restricted the export of these systems to a small set of close allies, as indicated in Table 5.1. Israel, although not a signatory of the MTCR, also abides by the category I restrictions and has been restrained in exporting these systems. Although Table 2.2 in Chapter Two shows only one export of category I UAVs from the UAE and none from China, this is likely due to a combination of two factors because they are not signatories of the MTCR. First, China and the UAE have only recently developed category I UAVs for export, and second, there is a lack of transparency in the export activities of these two nations, so we do not know which nations might have placed orders for these systems. As we discussed in Chapter Two, both China and the UAE are now not only marketing category I systems for export but also offering coproduction by building factories for manufacturing large UAVs. Additionally, as Table 2.2 indicates, the United States has been, by far, the largest exporter of category I UAVs, potentially because of MTCR restrictions and technological and cost barriers associated with developing these large systems.

The United States is no longer the sole manufacturer of category I UAVs, and others, especially non-MTCR nations, are now able and willing to export these systems. China, and potentially the UAE in the future, are undermining the MTCR's effectiveness by not only marketing and exporting large UAVs but also offering coproduction agreements for these vehicles. Therefore, from a global and practical perspective, the MTCR's effectiveness with category I UAV nonproliferation is decaying, and we expect that proliferation will accelerate in the near future.

Other Considerations Regarding Missile Technology Control Regime Export Restrictions on and Proliferation of Unmanned Aerial Vehicles

Rejecting a UAV export request, regardless of the reason, likely results in some response by the requesting nation. Examples could include any of the following:

- Abandon the pursuit of large UAVs altogether; this will, of course, depend on the original motivation for the request.
- Seek other foreign providers, potentially non–MTCR signatories.

---

not permitted to work directly with the U.S. company manufacturing the system. For a discussion of the FMS process, see Security Assistance Management Manual, undated.

- Attempt to purchase, from U.S. or foreign sources, slightly smaller UAVs that are below the category I threshold but meet the purchaser's mission needs.
- Develop an indigenous capability for manufacturing large UAVs. The purchasing nation might also seek assistance from non–MTCR signatories to develop this capability.
- Consider the lease of contractor-owned and -operated U.S. systems, although this approach also triggers an MTCR review.
- Convert manned aircraft into unmanned. This approach would be appealing to a small set of missions that do not require long endurance (e.g., delivery of munitions to target or use to overwhelm air defenses).
- Obtain and operate large UAVs through an alliance in which the United States participates.

In most cases of denial (actual or presumed), we found that nations reacted by purchasing large UAVs (category I or near–category I) from other sources, typically China. This course of action applies to the UAE, Jordan, Saudi Arabia, and others. Some of the requesters that were denied U.S. systems, such as the UAE and Saudi Arabia, pursued indigenous development programs. Others might also have chosen to convert some military aircraft into unmanned versions. The final option we provided in the list involves operating large UAVs through an alliance (e.g., NATO is obtaining several Global Hawk UAVs, as indicated in Table 5.1).

An additional consideration is the evolving large commercial UAV market. Although it is still in the early development phase, we assess, based on industry feedback and current reports, that the use of category I–sized UAVs in future commercial applications (e.g., freight delivery) is very likely. Commercial applications of large UAVs will likely further proliferate these systems. Integration of UAVs into unrestricted parts of the National Airspace System (NAS) will need to be approved before this happens, but, as discussed in previous chapters, significant progress is being made in that area. Another concern is the advent of OPA, as discussed earlier in the report: Germany is codeveloping the Q01. Another application of OPA is to help reduce the crew size in commercial airliners. Although completely eliminating the flight crew will be technically achievable, a potentially more likely approach would be to reduce the number of pilots needed to fly a commercial airliner to just one (Falk, 2017). This could require adding a remote-piloting capability from the ground in case of emergencies (i.e., if the pilot is incapacitated). The ground controller would be able to control and land the plane in these emergencies, which effectively makes the aircraft an OPA.

Rejected Export Requests

As mentioned in the previous section, the United States can deny an export request simply by not responding to the request. Also, the strong presumption of denial associated with category I UAVs effectively inhibits non-MTCR nations from initiating export requests from the United States. Nations that have been tacitly refused U.S. UAV purchase include Middle Eastern countries that could have been denied for other reasons, such as the UAS export policy or the policy of a qualitative military edge put in place to help Israel maintain military superiority in the Middle East (Scarborough, 2015). However, some of the interested nations that have been denied export include MTCR signatories, such as New Zealand and Brazil, and nonsignatories, such as Indonesia and Nigeria.[6]

Category II Exports

The MTCR is relatively flexible with respect to category II exports.[7] A case-by-case review of export applications is required (led by the Department of State), with the review examining proliferation and terrorism concerns. We note that the near–category I UAVs we previously introduced are category II under MTCR classification. There is a large and active export market for category II systems. UAV manufacturers have modified category I UAVs by reducing their payload capacities to convert them into category II systems. As an example, General Atomics has reduced the payload capacity for one system type below 500 kg and is marketing the system as an XP (for export) version.

Current U.S. Commercial Exports

As of 2013, the Department of Commerce is responsible for approving all commercial UAV exports. The number and value of approved category II commercial export applications has grown rapidly, reaching $47 million in the first half of the 2017 calendar year (U.S. Department of Commerce, 2017). However, there have been no category I export applications yet.

Summary

We assess that the MTCR's strong presumption of denial associated with category I has contributed to limiting the proliferation of these systems in the past. The United States has exported a significant number of category I UAVs to a small set of close allies, and most MTCR signatories have so far refrained from developing these large vehicles. However, the environment is rapidly changing as a capable global large-UAV industry is evolving. Non-MTCR nations, especially China, have been widely exporting near–category I systems, and China not only has recently been marketing a military category I UAV for export but is also building a manufacturing facility in Saudi Arabia to coproduce large UAVs, potentially including the category I CH-5. Other nations, such as the UAE and, to a lesser extent, Germany and potentially Italy, are also starting to develop and export category I UAVs. The strict strong presumption of denial for category I UAVs could hinder the development of future large U.S. commercial UAVs and might inhibit U.S. companies from exporting these commercial-purposed systems in the future.

---

[6] UAV industry representatives, discussions with the authors, Washington, D.C., July 26, 2017 (name withheld by agreement).

[7] A category II RPA is defined as having a range capability of at least 300 km but a payload capability below 500 kg. The MTCR does not cover RPAs with range capabilities below 300 km unless they are intended for WMD delivery. Any rocket or UAV intended for WMD delivery is treated at least as restrictively as category I.

CHAPTER SIX

Benefits and Risks of Limiting Exports of Unmanned Aerial Vehicles

This chapter addresses the fifth request in Section 1276 of the 2017 NDAA: "an assessment of the benefits and risks of continuing to limit exports of such aircraft."

In this chapter, we evaluate the potential security, economic, and political consequences of the United States maintaining the controls of category I UAV exports.[1]

Although the MTCR restrictions influence export decisions, they are only one of many U.S. control mechanisms to which an export request is subjected—others include the Wassenaar Arrangement on Export Controls for Conventional Arms and Dual-Use Goods and Technologies, the policy of a qualitative military edge, and the recently established UAS policy (Sharp, 2012). The latter is a new policy designed specifically for U.S.-made military and commercial UAS. It governs the international sale, transfer, and subsequent use of U.S. UASs and supplements and builds on the U.S. Conventional Arms Transfer Policy (U.S. Department of State, 2015). Therefore, like any other weapon system (and related technology), category I UAVs will be subjected to export-control considerations regardless of what MTCR restrictions are in place.

To assess the relative benefits and risks of the United States continuing to limit exports of UAVs, we identified evaluation parameters in three categories: security, economic, and political considerations. In the sections that follow, we define the parameters and discuss how they are affected by the category I export restrictions. A positive effect for a particular characteristic implies a beneficial outcome to the United States, while a negative mark implies an undesired consequence. A neutral assignment can result from either a no-effect assessment or from a combination of equally positive and negative impacts. The order in which we present and discuss the different categories and factors does not imply priority. Although many readers would consider security the most important category, we do not assign a priority level to any of the factors we review. The assessments are qualitative and were performed by RAND subject-matter experts using publicly available information and their expertise in the subject areas. The narrative in the assessment offers an objective discussion that accounts for both the positive and negative impacts of the United States limiting the export of category I UAVs. We have high confidence in the discussions presented but admit that the summary assessment provided for each factor is subject to variation depending on the specific circumstances a reader might be considering. As an example, in the "Cost to U.S. Customers" section of this chapter, we assert that the cost of manufacturing decreases as the number of units increases, and, because export

---

[1] This chapter addresses the impact of the MTCR export control of category I UAVs. We do not address the near–category I subset because they are considered category II in the MTCR and are not subject to the strong presumption of denial.

controls generally limit the number of units manufactured, the cost to U.S. customers should be higher. However, we assessed the impact as neutral because other factors that are difficult to assess (e.g., manufacturer's pricing strategy) need to be taken into account and are important.

Security Factors

A variety of security factors should be considered in an assessment of the benefits and risks of continuing to limit exports of category I UAV systems. Assessing these security factors is complex, especially because the United States has shown that MTCR restrictions on exports to allies and partners do not always constrain it; if the United States does not provide a system to an ally or partner, that nation can frequently obtain a similar capability from another foreign manufacturer. The factors we examine in this section are as follows:

- threat to U.S. forces: whether the proliferation of UAV systems represents an increased threat to U.S. forces and the impact of category I UAV export controls on that threat
- capability of allies: the effect that UAV proliferation and export controls have on the capabilities of allies and partners
- interoperability: how the acquisition of UAV capability from U.S. or foreign producers affects interoperability with allies and partners
- operational expertise: whether proliferation of UAV systems affects U.S. operational expertise in employing UAVs
- protection of U.S. UAV technology: Limiting exports can affect the protection of sensitive U.S. UAV technologies.

The Threat to U.S. Forces from Proliferation of Unmanned Aerial Vehicles

The growing proliferation of UAV systems represents an incremental increase in overall threat to U.S. forces. UAVs provide adversaries with a relatively low-cost means of conducting long-range surveillance of U.S. forces, especially during the preconflict phase, when UAVs can be used to locate, track, target, and (if the UAVs are armed) potentially launch preemptive strikes against U.S. ships, bases, or ground forces. Adversary employment of large UAVs is less of a concern during conventional conflict, when U.S. air defense systems will likely provide effective protection against most UAVs, but increased employment of these systems will complicate air defense efforts at a time when those defenses must now also deal with increased threats from cruise missiles.

As discussed in Chapter Two of this report, indigenous development and sales of category I (and sophisticated near–category I) UAV systems by and from foreign suppliers are increasing despite (or perhaps because of) a restrictive U.S. export policy, almost ensuring that U.S. forces will confront adversaries operating highly capable UAVs in any conflict or confrontation regardless of current or future export-control policy. Because of the recent availability of advanced UAV systems from non-U.S. manufacturers, U.S. export-control policy now has only a limited effect on the increasing threat from UAV proliferation (especially to potential U.S. opponents, because the United States would not export these systems regardless of MTCR restrictions). We therefore assess the effects that export control has on the threat to U.S. forces to be neutral.

The Capability of U.S. Allies and Partners

We judge that the export of U.S. category I UAV systems has increased the capability of U.S. allies and partners, which directly benefits the United States. Several allies and partners have employed their category I UAVs in support of counterterrorism operations in Afghanistan, while others are acquiring systems to enhance their capabilities to conduct ISR operations in critical areas, such as the Korean peninsula and the NATO area of operations (Ross, 2013; "UK Drone Strike Stats," 2017; Cole, 2016).

Export controls on category I UAVs have limited some allies' and partners' ability to acquire U.S. systems, thus reducing their potential capabilities and coalition contributions. Interviews with industry representatives and available data revealed that, although the United States has exported category I UAVs to allies and some MTCR signatories, the approval process can inhibit transactions, and partner nations that are not formal treaty allies or are outside of the MTCR are generally precluded from purchasing category I U.S. systems (Brannen, 2015; Berger, 2015). Thus, some potential partners cannot acquire U.S. systems that could be used in coalition operations, or they are acquiring less capable and less interoperable systems from other countries. We therefore assess the effects that export control has on the capabilities of U.S. allies and partners to be negative.

Interoperability with U.S. Allies and Partners

Interoperability involves primarily the ability to integrate a UAV system into the C2 architecture for safe and responsive operation of the system in a joint operating area and effective and timely sharing of any information that the UAV collects across coalition networks. Because of the different configurations of systems sold and variations in network architectures, interoperability with partner UAV systems is challenging whether the allies or partners operate U.S. systems or systems provided by other countries. As discussed in Chapter Four, interoperability is less problematic with U.S. systems, and it is much more likely that a U.S. system operated by an ally or partner will achieve a higher level of interoperability with U.S. forces (and enable future closer integration with U.S. systems) than systems from a foreign manufacturer will.[2] We conclude that current export restrictions on category I UAVs have a somewhat negative impact on interoperability between U.S. and allied forces.

Operational Expertise with Unmanned Aerial Vehicle Systems

Military planners, equipment operators, and manufacturers all benefit from the opportunity to operate their systems. The operational experience they receive provides feedback on system performance that is normally used to improve procedures and make modifications to platforms and supporting architectures. Permitting wider exports of U.S. UAVs would increase the amount of operational expertise available to the U.S. UAV community while precluding adversaries from acquiring comparable levels of operational experience when a country purchases a U.S. system rather than one from an adversary nation. Additionally, allies and partners can leverage UAV operational expertise resident in U.S. forces to improve their own operational capabilities. We judge that current export restrictions have a somewhat negative effect on U.S. and allied operational capabilities.

---

[2] See Chapter Four for a more detailed discussion of interoperability.

Protection of U.S. Unmanned Aerial Vehicle Technology

In general, increasing exports of any weapon system increases the risk that technology in that system will be compromised either by the purchasing nation or a third party. According to military subject-matter experts involved in U.S. security assistance programs, technology risk can be a significant concern when providing UAV systems to some U.S. allies and partners. Many nations do not apply the same level of operational security to their weapon systems as the United States does, and sometimes they transfer these systems to third parties.[3] The lost technology could be used to develop countermeasures against U.S. platforms or to enhance the capabilities of another manufacturer's systems. Logically, the risk of illicit technology transfer will increase the more systems are sold and the greater the number of nations that acquire these capabilities from the United States. However, we also note that the United States has exported billions of dollars' worth of advanced weapons in the past and has extensive experience in limiting the loss of its weapon technologies. Thus, we conclude that current export control on category I UAV systems reduces the risk of U.S. technology being compromised and therefore assess it as positive.

Economic Factors

UAV export controls have a variety of economic impacts on U.S. industry. We evaluate these impacts examining the following factors:

- global market share of category I UAVs: the U.S. fraction of total global sales
- cost to U.S. customers: the UAV unit price that U.S. customers would pay
- industrial base: the manufacturers, as well as suppliers and subcontractors involved in developing, manufacturing, and supporting the UAV industry
- research and development (R&D): the financial investment allocated to UAV technology R&D
- future opportunities: share of potential future commercial and civilian market opportunities related to large-UAV applications.

Global Market Share of Category I Unmanned Aerial Vehicles

The United States is currently the largest exporter of category I UAVs; however, qualitative information obtained from discussions with industry representatives indicates that U.S. industry has lost a significant number of category I UAV export opportunities as a result of export-control restrictions. Several countries that could not purchase U.S. systems turned to foreign sources to fulfill their needs.[4] Although it is unclear what fraction of the export rejections the MTCR caused, nations that have been denied UAV exports, such as Saudi Arabia, were approved for other U.S. advanced weapon systems (Muralidharan, 2017). As discussed in Chapter Two, foreign UAV manufacturers are in the process of developing category I systems

---

[3] Foreign desk officer stationed at a U.S. embassy overseas, telephone communication with the author, September 6, 2017 (name withheld by agreement).

[4] Industry representatives stated that Jordan, the UAE, and other countries indicated interest in U.S. category I systems but ended up purchasing foreign systems after the United States did not respond to their repeated requests (UAV industry representatives, interview with the authors, California, August 8, 2017b [name withheld by agreement]).

or have already done so and are offering them for export. Therefore, the United States is losing market share to those nations willing to bypass MTCR constraints and more easily approve the export of category I UAVs to allies and partners to which the United States has opted not to sell these systems. Given the available information, we assess that maintaining category I restrictions reduces the U.S. market share of global UAV sales and therefore assess it as negative.

Cost to U.S. Customers

Typical cost curves exhibit a unit price reduction as a larger number of a given product is manufactured. This is especially true for very large numbers of units (e.g., many thousands), at which point economies of scale apply. A generally accepted rule of thumb is that labor cost associated with aircraft manufacturing is reduced by 10 to 15 percent as the number built is doubled (Handy, 2013). Additionally, costs associated with nonrecurring expenses, such as R&D investments and manufacturing tools, are spread among a larger number of units sold. Thus, relaxing MTCR category I restrictions might result in more of these UAVs being built, which would likely reduce the average cost of a category I UAV system.

However, it is unclear whether a reduction in unit cost to manufacture UAVs would be passed on to U.S. buyers. Actual prices that manufacturers charge for high-value products, such as UAVs, involve complex considerations that go beyond just manufacturing costs and are proprietary. In addition, although relaxing category I UAV export control would likely result in more category I UAV exports, these purchases could replace some category II orders. A nation might order fewer category II UAVs if it purchases some category I systems and thus influence a manufacture's overall sales and pricing of future systems. The opposite is also possible (i.e., maintaining the category I restrictions could, in some cases, motivate some nations to purchase category II systems instead).

Overall, although we conclude that current export controls on UAVs can cause a higher unit cost for U.S. customers, we cannot assess whether those cost savings would actually be realized. Therefore, lacking supporting information, we assess this factor to be neutral.

Industrial Base

An increase in UAV manufacturing will help maintain and grow the UAV industrial base in the United States. Conversely, a flat or decreased rate of manufacturing will likely result in the eventual shrinking of the supporting industrial base. Although we are not aware of a comprehensive study evaluating the impact that export control has on the U.S. UAV industry, we can examine a closely related issue: the impact that export control has on satellite manufacturers in the U.S. space industry. The U.S. Department of Commerce recently completed a deep-dive assessment evaluating the effects that export restrictions on space-related technologies and products have on U.S. industry. It found that export controls had significant adverse effects on the industry, ranging from lost sales opportunities to contributing to the development of a capable and competitive foreign space industry (Botwin, 2014). We expect that export controls will have similar, if not more extensive, effects on the UAV industry. Therefore, we assess that current controls have a negative effect on the U.S. industrial base.

Future Opportunities

Determining future opportunities for a current product involves several uncertainties, including continued need of the benefit that the product provides and the possible development of disruptive or alternative technologies that might better fulfill user needs. However, we expect

UAVs of all sizes to be more widely used, provide more services, and replace and augment manned aircraft. Current studies predict continued growth in UAV spending, with an increase of more than 40 percent over the next five years (2017–2022) (Federal Aviation Administration [FAA], undated, p. 65). In 2009, FAA created the Unmanned Aircraft Program Office to integrate UAS into the NAS. FAA believes that the civil UAS markets will evolve and estimates that roughly 7,500 commercial UASs would be viable five years after integration of UAVs into the NAS (FAA, undated, p. 65). Companies in China and the UK have been planning and working on cargo UAVs. Although the initial ones are advertised to not exceed the category I threshold, they provide a clear path for larger, more-capable commercial cargo UAVs that exceed the MTCR limits.[5]

General Atomics has been actively developing systems to enable the integration of UAVs into the NAS and is scheduled to deliver the certifiable Predator B to the UK in 2019 (Carey, 2017). It meets NATO and UK certification requirements and is expected to be able to fly in civilian airspace (Carey, 2017). Other nations are considering the integration of UAVs into their NASs, and initial tests flying civil UASs in unrestricted airspace have already taken place in Europe (Antunes, 2016). U.S. industry representatives indicated that the current MTCR export restrictions will be a serious impediment to an anticipated future commercial autonomous-freighter business. The MTCR also excludes codevelopment of category I UASs resulting in potential opportunity losses in both commercial and military systems.

We assess that current export policy for category I systems has a negative impact on future commercial and military opportunities. However, we cannot quantify the level of loss, so we rate this factor as somewhat negative.

Research and Development

Sufficient R&D investments are needed to help maintain the lead position and future competitiveness of U.S. industry. Export controls can have several consequences for R&D efforts, potentially restricting investment for three reasons:

- R&D funding drops simply due to smaller budget availability. R&D expenditures are typically based on a percentage of company product sales; smaller sales result in less funding for R&D.
- Companies and other organizations, such as universities, will invest less R&D into products they cannot export (Botwin, 2014, p. 36).
- The amount of R&D in which companies invest relates to the expected payback (i.e., market size for future products being considered). A smaller market—U.S. only—will naturally result in smaller R&D investments.

The conclusions of the space study, from an R&D perspective, are consistent with the feedback we received from UAV industry representatives. Thus, we conclude that export controls could be reducing R&D investment.

---

[5] For near-term systems, see Thomson, 2017, and "UK Team Working to Development Mid-Mass Logistics Drone," undated. For longer-term plans, see Lennane, 2015.

Political and International Factors

Political and international factors must be considered when assessing the effects of category I UAV export controls. The factors we examine in this section are as follows:

- partner relations: Providing category I UAVs affects the relationship with the importer of those systems. Conversely, the rejection of the request can also affect that relationship.
- misuse of UAV systems: Nations acquiring UAVs can use them inconsistently with international law or U.S. values.
- arms control and MTCR implications: The United States exporting category I UAVs affects arms-control agreements and the behavior of other MTCR signatories.
- international motivation to develop UAVs: Additional nations have incentive to develop category I UAVs if category I MTCR controls are relaxed.
- global proliferation: Relaxing category I restrictions results in the spread of category I UAVs throughout the world.

Partner Relations

As discussed in Chapter Five, the United States has approved export of a significant number of these systems to allies and partners (mostly MTCR signatories); however, the process involved in the acquisition is time-consuming and opaque and can result in unpredictable outcomes. During discussions with industry representatives, it was noted that even close allies can have their requests rejected.[6] Other U.S. partners, such as Jordan, were also denied U.S. systems, sometimes driving them to acquire similar systems from foreign sources, such as China and, in some cases, Israel (Page and Sonne, 2017).

Although partner relationships are multidimensional, involving many factors, denying exports of category I UAVs can cause some frustration for the partner and drive it to pursue indigenous production programs or establish relationships with other nations that will provide these systems for them (Page and Sonne, 2017). Other consequences of MTCR controls include the implicit inability to partner with allies and others for joint development or manufacturing of category I systems. We conclude that current controls have a negative impact on partner relationships because they can, in some cases, result in U.S. partners developing military relationships with U.S. competitors.

Misuse of Unmanned Aerial Vehicle Systems

The international community is very concerned about the misuse of UAV systems. Potential abuses include improper surveillance, covert strike operations, and violations of human rights, such as extrajudicial killings (Kreps, 2014). The proliferation of UAV systems in the absence of controls or specific global norms that address their use and misuse creates a perception that suppliers are ignoring or even enabling the potential for abuse.

Partners that operate U.S. UAV systems offer both a risk and an opportunity in this area. The risk is that the partner will misuse those systems, especially for lethal operations, opening the United States to criticism for supplying the systems and promoting violations of human

---

[6] An industry representative indicated that New Zealand requested a category I UAV and the U.S. government rejected the request. However, further discussions with a U.S. government representative indicated that the export request is being reconsidered.

rights. On the other hand, U.S. weapon systems include end-use controls, and the United States can sometimes attach additional conditions on use of systems as part of the export agreement in an effort to prevent such misuse (Sayler et al., 2016).

Although most allies and partners would likely abide by U.S. end-user agreements, as more countries operate more systems, the probability of misuse will increase. Moreover, many smaller and foreign UAVs that are not subject to export controls can also perform undesired activities—so export controls alone are insufficient in preventing misuse. Moreover, long-endurance manned platforms that are optimized for ISR and irregular-warfare missions, such as the U.S. MC-12 or Brazilian EMB 314, can also, in principle, be misused in similar ways. Although we discuss misuse related to both nonlethal (e.g., surveillance operations) and lethal (e.g., extrajudicial killings) use, the latter, of course, gets substantially more attention from the public.

Looking at the totality of this factor, we assess that current restrictions on category I exports somewhat decrease the potential for misuse of UAV systems and therefore assess them as somewhat positive.

Implications of the Missile Technology Control Regime and Nonproliferation Agreements

MTCR signatories tend to follow the U.S. lead with regard to export control. If the United States unilaterally relaxes its policy regarding the export of category I UAVs, other signatories will likely follow suit. If the United States and other MTCR signatories choose to ignore or circumvent provisions of the regime, or appear to do so, it would likely be detrimental to the MTCR and could also have a negative effect on other multilateral nonproliferation agreements. The reach of this effect is difficult to predict or even measure, but potential consequences include the overall weakening of the MTCR. The MTCR has been successful in limiting the proliferation of large missiles, and we expect that it would continue to be effective in that arena independently of UAV policies.

We assess that maintaining the status quo with respect to U.S. export controls has a positive impact on the MTCR and other nonproliferation agreements.

International Motivation to Develop Unmanned Aerial Vehicles

Changing the export controls on category I UAVs is not likely to have a significant impact on other countries' decisions to develop UAVs. Relaxing the MTCR restrictions will not directly affect the non-MTCR nations that are already manufacturing category I UAVs. In theory, states that are pursuing UAV development because they cannot buy U.S. systems could abandon those efforts if they could obtain U.S. UAVs. A nation with significant UAV development and manufacturing capabilities will likely not give those up as a result of the United States relaxing its export control. However, countries with only rudimentary capabilities might decide to abandon their efforts and acquire U.S. systems instead. MTCR signatories currently able to manufacture large UAVs might be motivated to export them if they believed that there were a growing export market; however, they would also struggle to effectively compete with capable and established exporters, such as Israel, the United States, and China. Overall, we conclude that current MTCR restrictions have motivated foreign development of large UAVs. Indeed, the increasing UAV development and manufacturing capabilities in China and other nations can be partially attributed to U.S. export controls (Chuanren and Pocock, 2017).

Global Proliferation

Proliferation of UAV systems has increased substantially in the past decade and is likely to continue (FAA, undated, p. 65). Most of the recent proliferation has been of category II and smaller systems; however, recent foreign marketing efforts to sell category I systems, as well as the inspiration for some foreign companies to coproduce large UAVs, indicate that nations with security requirements for category I systems and the resources to develop or acquire those systems will do so. We assess that current MTCR controls have a minor effect on controlling current and future proliferation, especially for nonsignatory nations. In the future, commercial UAV applications, such as cargo transport, will likely evolve and increase the demand for large UAVs—further motivating non-MTCR countries to increase their development and sales of these vehicles.

If the UAV category I restrictions are relaxed, it could result in a faster global proliferation of UAV systems with large payload capacity. Although allies and partners would have easier access to U.S. systems, some MTCR signatories that have so far avoided developing or exporting these systems might be motivated to start doing so.

In addition to the fact that capable near–category I systems are widely proliferated, the current willingness of non-MTCR nations to export category I UAVs (such as China openly exporting the CH-5) or by misclassification (e.g., Wing Loong II), the MTCR's effectiveness in limiting global proliferation of these UAVs is being seriously compromised. Considering the current and evolving UAV export market, we assess this factor to be neutral.

Summary of Our Assessments

We assessed the benefits and risks of limiting the export of category I UAVs (as requested by the NDAA) by identifying the security, economic, and political factors that these export limits could affect. We summarize these factors and their assessments in Table 6.1. The table provides an overall visual representation of our assessment and is supported by the discussions in the chapter. The table shows that the MTCR's effects on security considerations are somewhat negative. Although this might appear counterintuitive because tight export controls are limiting U.S. systems from proliferating, U.S. allies and partners are being negatively affected while the threat to U.S. and allied troops from foreign-made UAVs, mostly China, has increased. As

Table 6.1
Estimated Risks and Benefits of Continuing to Limit Exports of Category I Unmanned Aerial Vehicles

Impact	Positive	Somewhat Positive	Neutral	Somewhat Negative	Negative
Security		Protection of U.S. technology	Threat to U.S. forces	Interoperability; operational expertise	Allies' capabilities
Economic			Cost to U.S. customers	Future opportunities	Global market share; U.S. industrial base; R&D
Political	MTCR implications	Misuse of UAVs	Global proliferation	International motivation to develop UAVs	Partner relations

expected, the economic impact on the United States is negative because fewer sales are occurring. The political results illustrated in the table would appear to be neutral. However, we should also consider more closely the relative risk and consequence of each factor (e.g., "Partner relationship" versus "Misuse of UAVs") especially when assessing the political implications.

Looking across the assessments presented in the chapter, we conclude that the MTCR's effectiveness in controlling large UAVs has eroded for several reasons:

- A subset of category II systems, which we describe and label near–category I, are widely proliferated. These systems have capabilities that are near those of category I systems but with smaller payloads (between 300 and 500 kg).
- Several nations are now developing and openly marketing category I systems for sale to both MTCR and non-MTCR nations.
- Non-MTCR nations with the capability to manufacture category I systems (e.g., China) have also marketed these systems and, in one case, are building coproduction facilities to produce them.

CHAPTER SEVEN

Summary and Conclusions

UAVs have proliferated and continue to proliferate on a global scale. Civil institutions, such as the U.S. Department of Interior and FAA, are becoming prolific users of UAVs because of their numerous advantages over manned systems (Association for Unmanned Vehicle Systems International, 2017). Many foreign militaries now have some type of UAV and plan to further expand their arsenals. Publicly available data show that near–category I UAVs have proliferated, as shown in Figure 2.6 in Chapter Two. And although category I systems have not proliferated to the same extent, recent announcements indicate that nations that are motivated and can afford to acquire them will likely be able to do so from foreign sources—primarily China and the UAE in the near future.

Large UAVs are currently being used solely for military missions; however, future commercial applications of large UAVs are starting to surface and are likely to grow. The UAV market is expected to grow at a significant rate, from about $6 billion in 2015 to more than $12 billion per year by 2025, and large aviation leasing companies are now including UAVs in their portfolios (Stohl, 2015). UAV development barriers have limited the number of nations willing to develop these systems. However, the information and data presented in the report point to an acceleration in nations developing large UAVs indigenously or with foreign assistance.

Overall, we find that the current pervasiveness of near–category I UAVs and the apparent future trend in the spread of category I systems lead us to conclude that a proliferation of large UAVs is ongoing and accelerating (Chuter, 2017).

Findings and Assessments of Missile Technology Control Regime Effects

Our analysis suggests that the MTCR has had some effect on controlling the proliferation of category I UAVs. The United States has limited the export of these vehicles to a small set of allies, and most MTCR signatories have so far restrained from developing these large UAVs. None, to our knowledge, has yet exported them to non–MTCR signatories. However, the landscape has dramatically changed in recent years. Non-MTCR nations, primarily China and, to a lesser extent, Israel, have exported large armed and unarmed near–category I UAVs. These systems do not technically qualify as category I, but they have very similar attributes and mission capabilities that are valuable to a large number of countries.

More significantly, China has recently openly marketed an armed category I system, the CH-5, and has reached significant deals to comanufacture this system and others. China is building a UAV factory in Saudi Arabia to develop up to 300 large UAVs, including possibly the CH-5. This deal could be a prelude of things to come. Some non-MTCR nations have shown interest in building up their UAV capabilities and might either purchase these large UAVs or decide to coproduce them with China's help. These types of arrangements not only increase military cooperation between the acquiring nation and China but also can influence collaboration with the United States. Furthermore, some cracks appear to be forming within MTCR signatories. Germany is codeveloping a category I OPA with Qatar, and Italy plans to export a potentially category I system to the UAE.

Although there are no current commercial category I UAVs, a Chinese company is developing a large cargo-delivery UAV, while U.S. industry representatives indicate that "the MTCR export restrictions will be a serious impediment to an anticipated future commercial autonomous freighter business" (Meredith and Kharpal, 2017).

Restricting large-UAV exports to allies and partners has several detrimental effects, including further complicating interoperability capabilities between the United States and its allies and partners in joint operations.

Although the MTCR has been, and continues to be, an effective tool to control the spread of ballistic and cruise missiles, the availability of these vehicles from non-MTCR nations has significantly eroded the MTCR's efficacy to limit the proliferation of large UAVs. Furthermore, the strong presumption of denial associated with category I UAVs could be affecting current U.S. industry plans and investments in future large commercial UAVs.

The report addresses the first five assessment areas posed in the NDAA:

1. We provided a qualitative and quantitative assessment of the scope and scale of the proliferation of RPAs that are category I items under the MTCR and showed that proliferation of these systems is accelerating. The assessment also included a subset of category II systems that we labeled as near–category I.
2. We assessed that the threat posed to U.S. interests as a result of the proliferation of such aircraft to adversaries varies depending on the conflict phase and the missions considered. We also noted that the near–category I systems can pose a threat to U.S. forces.
3. We assessed that the proliferation of the category I UAVs can improve the combat capabilities of U.S. allies and partners. However, we noted that the import of U.S. UAV systems, instead of those manufactured by potential U.S. opponents, could improve future interoperability with allies and partners.
4. We concluded that, although the United States has limited the proliferation of category I UAVs as a result of the application of a strong presumption of denial for exports of such aircraft, many of the nations denied these U.S. systems were able to acquire them through other means, including imports from and coproduction with China.
5. This report provides an assessment of the benefits and risks of continuing to limit exports of such aircraft by considering various security, economic, and political factors.

Open Questions, Future Analyses

The assessments provided in this report are based on published data and information, as well as discussions with different government and industry representatives. We did not get the opportunity to meet with representatives of foreign nations that the U.S.-enforced MTCR export controls affect. Such meetings would provide a direct assessment of the impacts and consequences of UAV export restrictions.

We also avoided making recommendations about what actions the United States could pursue to alleviate some of the negative impacts that the MTCR export control is causing. Although we discussed some of these issues with both government and industry representatives, such recommendations would require further analyses to assess their effectiveness and potential unintended consequences. Another factor we did not pursue is the availability of technologies to allow a country to build a category I UAV indigenously. The MTCR's effectiveness in preventing nations from acquiring category I components and indigenously building the aircraft is not clear and should be analyzed. Finally, a more in-depth analysis of the implications that the MTCR category I has on the evolving commercial UAV industry would provide a more comprehensive assessment of the potential risks that the category I restrictions pose for the U.S. industrial base.

Assessment of Missile Technology Control Regime Category I Restriction

This appendix discusses the potential UAV operational range limitations that a communication system causes. We also discuss a UAV's range and payload capacity, along with the capacities' effects on the MTCR restrictions.

The Extendibility of Unmanned Aerial Vehicles' Range and Communication Links

Range is one of two factors that determine whether the MTCR governs the exports of a given UAV platform as a category I technology. Understanding how UAV producers and export review committees define the range metric is important, as is assessing how the range restriction applies to U.S. and foreign UAV exports. The MTCR considers the range to be the maximum one-way distance that the UAV system can travel in stable flight over the earth's surface. It is based on the maximum capabilities of the design and is independent of external factors, such as operational restrictions or limitations imposed by telemetry or communication links (MTCR, 2010, p. xxii).

The MTCR definition of range is critical because the ranges that UAV producers give publicly have changed as technology has developed. More than 15 years ago, large UAVs with an endurance of dozens of hours were often described as having a range of only several hundred kilometers (e.g., Heron by IAI) (IAI, undated [a]) because of common operational limits in the line-of-sight (LOS) radio communications, including the telemetry downlink and the C2 uplink. In recent years, however, other communication technologies have become increasingly accessible to UAV manufacturers and exporters. Satellite communications (SATCOM) have become inexpensive, reliable, and common in even moderately advanced UAV systems (Davis et al., 2014).

Table A.1 shows the typical methods of C2 communication, approximate limits to operational range due to the communication package, and the difficulty of implementing that technology relative to implementing a single GCS LOS communication link. The values given in the table are, by necessity, very simplified and do not necessarily consider every possible technology, tactic, or scenario.

Ground-based LOS refers to the typical radio-frequency communications between the UAV platform and the primary GCS. This is the simplest communication scheme but also the most limited in terms of tactical range. UAV antenna requirements are relatively minimal, and no additional communication infrastructure is required.

Table A.1
Communication Approaches for Unmanned Aerial Vehicles

Communication Method	Typical Range Limit, in Kilometers	Ease of Development
Ground-based LOS	100–250	n/a
Grounded antenna relay	300 or greater	Easy
Airborne antenna relay	300 or greater	Moderate
SATCOM	Unlimited	Difficult (initially)
Full autonomy (no required communications)	Unlimited	Very difficult

NOTE: n/a = not applicable.

Grounded antenna relay indicates that one or more remote antennae receive and retransmit C2 communications beyond the LOS of the primary GCS. Most major UAV platforms (including exported Israeli platforms) have this as an option. It allows the extension of communication ranges without any modification to the UAV. However, it requires distributed communication infrastructure to be placed around the operational theater.

Airborne relays play a similar role but benefit from better LOS by using supporting UAVs close to the GCS as a platform for the antennae. Many major exported platforms have this as an option. This technique can reduce the influence that moderate terrain has on the LOS communication range. It allows the extension of communication ranges without any modification to the primary UAV but requires additional airframes and more-complicated communication equipment. The range extension for ground and airborne antenna relays depends on the communication infrastructure in place (e.g., the number and distribution of antenna relays).

SATCOM relay communication information, potentially anywhere on the planet, via a satellite relay. SATCOM have become inexpensive, reliable, and common among major UAV platforms in recent years. By using existing communication satellite infrastructure, a UAV can achieve effectively unlimited communication range. These methods are all common in exported UAVs by major producers, including the United States, Israel, and China.

Currently developing technologies also include a significant level of autonomous capability. *Full autonomy*, which eliminates the need for constant human monitoring of the UAV, could extend the range of reusable UAV platforms beyond any communication requirement. Although autonomous cruise missiles and loitering munitions are common (e.g., Elbit's Skystriker), this type of operation has generally been avoided in large reusable UAVs. In the future, increased sophistication in autonomous systems could reduce the need for constant communication between the GCS and UAV, and some analysts believe that this will also lead to increased range of UAV systems (Elbit Systems, undated [b]; Foust, 2013).

Most exported UAVs now have multiple options to extend their operations beyond LOS communications by using grounded antenna relays, airborne relays, and SATCOM. The previous limits to UAV ranges due to communication capabilities have largely been eliminated for a traditional state military.

Trading Unmanned Aerial Vehicle Range for Payload Capacity

Aircraft have a natural trade-off between range and payload. Several limiting factors contribute to the maximum payload capacity, including structural load limits, takeoff weight restrictions, and fuel consumption rate (a higher payload for a given aircraft requires higher fuel consumption). Large UAVs are traditionally designed to operate with very long endurances, but there is traditionally a trade-off between maximum endurance and useful payload capacity. For example, although the MQ-9 has a maximum endurance (unloaded) of about 27 hours, it has an endurance of only about 14 hours at maximum payload (General Atomics Aeronautical, undated [b]; Wheeler, 2012). This is primarily because the maximum takeoff weight of the MQ-9 (about 4,750 kg) does not allow both the maximum payload capacity (about 1,700 kg) and the maximum fuel load (about 1,800 kg) (U.S. Air Force, 2015).

Generally, the specified maximum payload and maximum endurance cannot be achieved in the same mission because the aircraft would be above the maximum takeoff weight. However, because of the long-endurance design intent of large UAVs, category I UAVs can generally achieve their maximum specified payload capacities and still have ranges well above 300 km.

A possible approach to increasing the payload capacity beyond the advertised limit is by reducing the UAV's endurance or range. For UAVs that are classified as near–category I in terms of capability (e.g., Wing Loong II or Heron TP-XP), the loaded fuel weight can be reduced substantially while still achieving a range that is above 300 km. For example, the MQ-9 (although already a category I) could achieve a range of 300 km with less than 100 kg of fuel, compared with a fuel load of about 800 kg while carrying the maximum specified payload. This leads to the danger that the payload can be increased in UAVs that have near–category I capability limits to be well above 500 kg. The extent to which this is possible without making major modifications to the UAV depends on other limitations, such as structural load limits and the availability of payload mounting points.

There are good reasons to believe that certain UAVs exported by other countries as category II items can achieve category I capabilities with minimal modification or by simply pushing a system beyond traditional aerospace engineering design safety factors. As an example, China's Wing Loong II (with a published payload capacity of 480 kg) could readily achieve a category I mission via moderate reductions in fuel load (while still reaching 300 km in range). Because payload mounting points are located near the center of gravity, adding payload will not significantly affect flight stability characteristics, and the small increase in payload (less than 50 kg) should be within the aircraft's loading capability if judiciously applied.

A similar example of a category II system that can attain category I classification is the Heron TP-XP. This is a version of IAI's most capable UAV (the Heron TP), which has been modified for export under the MTCR by reducing the published payload specification from 1,000 to 450 kg. The XP version of the Heron has long endurance, up to 40 hours, and we estimate that the aircraft can easily exceed the 300-km range while trading more than 50 kg of fuel for payload, which would make this UAV a category I system.

Additionally, the method used for converting the TP version to an XP version by reducing its useful payload capacity is unclear, but a review of the published specifications indicates that both versions have the exact same capability outside the payload weight and space. It is not clear how challenging it would be to reverse the modifications such that the platform would be capable of achieving near the original 1,000 kg of payload. We note that both the Heron TP-XP and Wing Loong II are considered near–category I for our study.

Export Data for Category I and Near–Category I Unmanned Aerial Vehicles

This appendix provides a basic description of category I and near–category I systems, including range and payload weight capacity. Tables B.1 and B.2 contain basic specifications of known category I and near–category I systems. Some of the near–category I systems are significantly less expensive than the larger category I, and we expect the lower cost to result in greater proliferation of these vehicles.[1]

---

[1] The near–category I Chinese CH-4 UAV is estimated to cost less than half what its larger category I CH-5 counterpart costs (see Chen, 2017; see also Chan, 2017).

Table B.1
Category I Unmanned Aerial Vehicle Platforms

Country of Manufacture	System	Range, in Kilometers	Payload, in Kilograms	Endurance, in Hours	Altitude, in Feet
China	Soar Dragon	7,000	650	10	59,055
	CH-5	250–2,000[a]	1,200	39, 60[b]	7,000
Israel	Heron TP	7,400	1,000	36	45,932
Spain	FLYOX I	5,064	1,850	25	24,000
UAE	Yabhon Smart Eye	n/a	550	120	23,950
	United 40 Block 1	n/a	1,000	120	22,966
United States	X-47B	3,889	2,041	6	40,000
	K-MAX	555	2,721	12	15,092
	Orion	24,140	1,179	120	30,000
	RQ-4A	22,224	907	35	60,000
	RQ-4B	22,779	1,361	36	60,000
	Predator C	n/a	2,984	20	50,000
	Predator B Reaper	8,519	1,746	32	50,000

SOURCES: "All the World's Aircraft," undated; "Elbit Hermes 900 (Kochav) Medium Altitude, Long Endurance (MALE) Unmanned Aerial Vehicle (UAV)," 2017; "Heron TP (Eitan) MALE UAV," undated; "Chang Hong-5 (CH-5) Combat and Reconnaissance Drone," undated.

NOTE: Ranges are as advertised.

[a] LOS data link and SATCOM data link.

[b] Petrol engine and heavy fuel engine.

Table B.2
Near–Category I Unmanned Aerial Vehicle Platforms

Country of Manufacture	System	Range, in Kilometers[a]	Payload, in Kilograms	Endurance, in Hours	Altitude, in Feet
China	CH-4	4,000	345	40	22,960
	Wing Loong II	2,000	480	32	29,520
Israel	Heron TP-XP	n/a	450	30	45,000
	Super Heron	1,000	450	45	30,000
	Hermes 900	2,500	350	36	30,000
	Dominator	300	376	20	n/a
India	Rustom	300	350	24	35,000
Italy	P.1HH HammerHead[b]	8,148	300[b]	16	45,000
Poland	ILX-27	450	300	n/a	13,120
United States	Centaur	1,296	363	24	27,500

SOURCES: "All the World's Aircraft," undated; Lappin, 2017; Piaggio Aerospace, undated; Aeronautics, undated.

[a] Ranges are as advertised; it is not always clear whether the range is consistent with the maximum payload.

[b] HammerHead payload varies by source (the manufacturer does not provide maximum payload) and could exceed 500 kg.

Bibliography

Aeronautics, "Dominator XP MALE UAS," fact sheet, undated.

"ALIT/CASC CH-4 Series," *Jane's Unmanned Aerial Vehicles and Targets*, February 14, 2017.

"All the World's Aircraft: Unmanned," *Jane's by IHS Markit*, undated. As of October 3, 2017:
http://janes.ihs.com/UnmannedAerial/Reference#

Antunes, Jose, "UAVs and Airplanes Flying in a Single European Sky," *Commercial UAV News*, June 7, 2016. As of August 14, 2017:
https://www.expouav.com/news/latest/uavs-airplanes-flying-single-european-sky/

Apthorp, Claire, "Certifiable Predator B: A New Generation Sharpens Its Claws," *Air Force Technology*, March 12, 2017. As of August 18, 2017:
http://www.airforce-technology.com/features/
featurecertifiable-predator-b-a-new-generation-sharpens-its-claws-5760856/

Armstrong, Ian, "What's Behind China's Big New Drone Deal?" *Diplomat*, April 20, 2017. As of August 24, 2017:
http://thediplomat.com/2017/04/whats-behind-chinas-big-new-drone-deal/

Association for Unmanned Vehicle Systems International, "Government Representatives Discuss Impact UAS During AUVSI's Hill Day Luncheon," September 12, 2017. As of September 12, 2017:
http://www.auvsi.org/industry-news/
government-representatives-discuss-impact-uas-during-auvsi%E2%80%99s-hill-day-luncheon

"AVIC Wing Loong Series," *Jane's Unmanned Aerial Vehicles and Targets*, March 20, 2017.

Batey, Angus, "Adcom Focuses on Global Expansion," *Aviation Week*, November 9, 2015. As of October 3, 2017:
http://aviationweek.com/dubai-air-show-2015/adcom-focuses-global-expansion

Berger, Judson, "Obama Administration Denied Predator Drone Request for Jordan, Rep Urges Reversal," *Fox News*, February 6, 2015. As of September 14, 2017:
http://www.foxnews.com/politics/2015/02/06/
obama-administration-denied-request-for-jordan-to-get-predator-drones-rep-urges.html

Biggers, Chris, "UAE United 40 Block 5 at Test Airfield," *Bellingcat*, March 26, 2015. As of December 16, 2017:
https://www.bellingcat.com/news/2015/03/26/uae-united-40-block-5-at-test-airfield/

Binnie, Jeremy, "General Atomics Confirms UAE Predator Delivery," *Jane's Defence Weekly*, February 17, 2017. As of August 14, 2017:
http://www.janes.com/article/67797/general-atomics-confirms-uae-predator-delivery

Blanchard, Christopher M., "Saudi Arabia: Background and U.S. Relations," Washington, D.C.: Congressional Research Service, RL33533, June 13, 2017.

Borys, Christian, "Ukrainian Forces Says Two Drones Shot Down over War Zone Are Russian," *Guardian*, May 21, 2015. As of October 3, 2017:
https://www.theguardian.com/world/2015/may/21/ukraine-drones-shot-down-russian

Botwin, Brad, *U.S. Space Industry "Deep Dive" Assessment: Impact of U.S. Export Controls on the Space Industrial Base*, Washington, D.C.: U.S. Department of Commerce, Bureau of Industry and Security, February 2014. As of December 14, 2017:
https://www.bis.doc.gov/index.php/documents/technology-evaluation/898-space-export-control-report/file

Bowen, Wyn Q., "U.S. Policy on Ballistic Missile Proliferation: The MTCR's First Decade (1987–1997)," *Nonproliferation Review*, Fall 1997, pp. 21–39. As of December 14, 2017:
https://www.nonproliferation.org/wp-content/uploads/npr/bowen51.pdf

Brannen, Kate, "U.S. Firm Denied Request to Market Drones to Jordan," *Foreign Policy*, February 5, 2015. As of October 3, 2017:
http://foreignpolicy.com/2015/02/05/u-s-firm-denied-request-to-market-drones-to-jordan/

Calderwood, Imogen, "Satellite Image Reveals China Has Begun Using Drones with Stealth Capabilities in the South China Sea," *Daily Mail*, May 27, 2016. As of September 1, 2017:
http://www.dailymail.co.uk/news/article-3613173/
Satellite-image-reveals-China-begun-using-drones-stealth-capabilities-South-China-Sea.html

Carey, Bill, "Thales, IAI Flight Test NATO-Standard Data Link on Heron," *AINonline*, July 8, 2015. As of August 10, 2017:
http://www.ainonline.com/aviation-news/defense/2015-07-08/
thales-iai-flight-test-nato-standard-data-link-heron

———, "New Multi-Mission Sky Guardian UAS Is More Than a Strike Drone," *AINonline*, June 14, 2017. As of December 14, 2017:
https://www.ainonline.com/aviation-news/defense/2017-06-14/
new-multi-mission-sky-guardian-uas-more-strike-drone

Chan, Minnie, "Chinese Drone Factory in Saudi Arabia First in Middle East," *South China Morning Post*, March 26, 2017. As of October 3, 2017:
http://www.scmp.com/news/china/diplomacy-defence/article/2081869/
chinese-drone-factory-saudi-arabia-first-middle-east

"Chang Hong-5 (CH-5) Combat and Reconnaissance Drone," *Global Security*, undated.

Chen, Stephen, "China Unveils Its Answer to US Reaper Drone: How Does It Compare?" *South China Morning Post*, July 17, 2017, updated July 18, 2017. As of August 29, 2017:
http://www.scmp.com/news/china/diplomacy-defence/article/2103005/
new-chinese-drone-overseas-buyers-rival-us-reaper

Chuanren, Chen, and Chris Pocock, "Saudi Arabia Buying and Building Chinese Armed Drones," *AINonline*, April 12, 2017. As of August 24, 2017:
http://www.ainonline.com/aviation-news/defense/2017-04-12/
saudi-arabia-buying-and-building-chinese-armed-drones

Chuter, Andrew, "Aviation Leasing Giant Enters the Fray of Remotely Piloted Air Systems," *Defense News*, September 12, 2017. As of September 16, 2017:
http://www.defensenews.com/digital-show-dailies/dsei/2017/09/12/
aviation-leasing-giant-enters-the-fray-of-remotely-piloted-air-systems/

Clark, Bryan, Peter Haynes, Jesse Sloman, and Timothy Walton, "Restoring American Seapower: A New Fleet Architecture for the United States Navy," Center for Strategic and Budgetary Assessments, February 9, 2017. As of December 14, 2017:
http://csbaonline.org/research/publications/
restoring-american-seapower-a-new-fleet-architecture-for-the-united-states-

Cohen, Gili, "German Lawmakers Block Israeli Drone Deal After Discovering They're Armed," *Haaretz*, June 29, 2017. As of August 8, 2017:
http://www.haaretz.com/israel-news/1.798475

Cole, Chris, "European Use of Military Drones Expanding," *Drone Wars UK*, July 19, 2016. As of September 14, 2017:
https://dronewars.net/2016/07/19/european-use-of-military-drones-expanding/

Dassault Aviation, "European MALE Drone Development: Airbus, Finmeccanica and Dassault Aviation Welcome the Signature of the Trinational Declaration of Intent by Germany, Italy and France," May 18, 2015. As of September 7, 2017:
https://www.dassault-aviation.com/en/group/press/press-kits/european-male-drone-development-airbus-finmeccanica-and-dassault-aviation-welcome-the-signature-of-the-trinational-declaration-of-intent-by-germany-italy-and-france/

Davis, Lynn E., Michael McNerney, James S. Chow, Thomas Hamilton, Sarah Harting, and Daniel Byman, *Armed and Dangerous? UAVs and U.S. Security*, Santa Monica, Calif.: RAND Corporation, RR-449-RC, 2014. As of December 14, 2017:
https://www.rand.org/pubs/research_reports/RR449.html

Defense Advanced Research Projects Agency, "Broad Agency Announcement: CONverged Collaborative Elements for RF Task Operations (CONCERTO)," April 2016.

Defense Security Cooperation Agency, "Major Arms Sales," undated. As of December 16, 2017:
http://www.dsca.mil/major-arms-sales

"Distributed Common Ground System (DCGS)," *C4ISR and Mission Systems: Joint and Common Equipment*, September 11, 2017.

Dominguez, Gabriel, "China's AR-2 UAV-Capable Air-to-Surface Missile Ready for Export, Says Report," *Jane's Defence Weekly*, February 6, 2017.

Drew, James, "Marines Want EW Payloads for Insitu 'MQ-21' Blackjack," *Aerospace Daily and Defense Report*, October 27, 2016. As of October 3, 2017:
http://aviationweek.com/awindefense/marines-want-ew-payloads-insitu-mq-21-blackjack

"Drone B-17," photograph, National Archives and Records Administration 342-FH-3A40606, undated. As of September 4, 2017:
https://airandspace.si.edu/multimedia-gallery/7377hjpg

Egozi, Arie, "Russia Developing Unmanned Forpost-M," *FlightGlobal*, March 21, 2017. As of October 3, 2017:
https://www.flightglobal.com/news/articles/russia-developing-unmanned-forpost-m-435419

"Elbit Hermes 900 (Kochav) Medium Altitude, Long Endurance (MALE) Unmanned Aerial Vehicle (UAV)," *Military Factory*, July 1, 2017. As of December 20, 2017:
https://www.militaryfactory.com/aircraft/detail.asp?aircraft_id=1236

Elbit Systems, "HermesTM Universal Ground Control Station (UGCS)," undated (a). As of August 10, 2017:
http://elbitsystems.com/uas-hermes-universal-ground-control-station-ugcs/

———, "Skystriker," undated (b). As of August 21, 2017:
http://elbitsystems.com/product/skystriker/

Emmott, Robin, "Italy, France, Germany Sign European Drone Project," Reuters, May 18, 2015. As of August 14, 2017:
http://www.reuters.com/article/us-eu-drones-idUSKBN0O312A20150518

Ephron, Dan, with Kevin Peraino, "Hizbullah's Worrisome Weapon; Imagine If Terrorists Got Hold of Car Bombs with Wings—Now They Can," *Newsweek*, September 11, 2006.

FAA—*See* Federal Aviation Administration.

Falk, Dan, "Self-Flying Planes May Arrive Sooner Than You Think—Here's Why," *MACH*, October 11, 2017. As of December 17, 2017:
https://www.nbcnews.com/mach/science/self-flying-planes-may-arrive-sooner-you-think-here-s-ncna809856

Federal Aviation Administration, *FAA Aerospace Forecast: Fiscal Years 2013–2033*, undated. As of December 14, 2017:
https://www.faa.gov/data_research/aviation/aerospace_forecasts/media/2013_Forecast.pdf

Fisher, Richard D., Jr., "Military Parade Reveals Turkmenistan's New Chinese-Built UAVs," *IHS Jane's Defence Weekly*, November 2, 2016.

Foreign desk officer stationed at a U.S. embassy overseas, telephone communication with the author, September 6, 2017 (name withheld by agreement).

Foust, Joshua, "Why America Wants Drones That Can Kill Without Humans," *Defense One*, October 8, 2013. As of August 21, 2017:
http://www.defenseone.com/technology/2013/10/ready-lethal-autonomous-robot-drones/71492/

"France: Air Force," *Jane's World Air Forces*, July 25, 2017.

Freedberg, Sydney J., Jr., "Marines Aim for Jammers on 'Every Airplane,'" *Breaking Defense*, April 1, 2016. As of October 3, 2017:
http://breakingdefense.com/2016/04/marines-ramp-up-cyber-electronic-warfare-jammers-on-every-airplane/

———, "Army Races to Rebuild Short-Range Air Defense: New Lasers, Vehicles, Units," *Breaking Defense*, February 21, 2017. As of September 1, 2017:
http://breakingdefense.com/2017/02/army-races-to-rebuild-short-range-air-defense-new-laser

Gady, Franz-Stefan, "Chine Scores Biggest Military Export Order for Killer Drones," *Diplomat*, March 2, 2017. As of August 14, 2017:
http://thediplomat.com/2017/03/china-scores-biggest-military-export-order-for-killer-drones/

GAO—*See* U.S. Government Accountability Office.

General Atomics Aeronautical, "Advanced Cockpit GCS," undated (a). As of August 10, 2017:
http://www.ga-asi.com/advanced-cockpit-gcs

———, "Predator B RPA," undated (b). As of August 21, 2017:
http://www.ga-asi.com/predator-b

Glaser, April, "One of China's Biggest Online Retailers Is Building a Delivery Drone That Can Carry 2,000 Pounds of Cargo," *Recode*, May 22, 2017. As of October 3, 2017:
https://www.recode.net/2017/5/22/15666446/jd-china-drone-delivery-two-thousand-2000-pounds

Handy, Jim, "Moore's Law vs. Wright's Law," *Forbes*, March 25, 2013. As of August 13, 2017:
https://www.forbes.com/sites/jimhandy/2013/03/25/moores-law-vs-wrights-law/

Haria, Rupa, "UAE UAV Eyes Global Market," *Aviation Week*, November 23, 2013. As of December 15, 2017:
http://aviationweek.com/blog/uae-uav-eyes-global-market

"Heron TP (Eitan) MALE UAV," *Air Force Technology*, undated. As of December 20, 2017:
http://www.airforce-technology.com/projects/heron-tp-eitan-male-uav/

Hu, Jing, J. B. Gao, F. L. Posner, Y. Zheng, and W. W. Tung, "Target Detection Within Sea Clutter: A Comparative Study by Fractal Scaling Analyses," *Fractals*, Vol. 14, No. 3, September 2006, pp. 187–204.

Hunter, Duncan D., "Jordan Needs U.S. Drones to Fight ISIS," *Wall Street Journal*, September 21, 2015, p. A13.

IAI—*See* Israel Aerospace Industries.

"IAI Heron," *Jane's Unmanned Aerial Vehicles and Targets*, May 22, 2017.

"India: Air Force," *Jane's World Air Forces*, July 21, 2017.

"India Deploys US-Made Surveillance Drones Along LoC," *Express Tribune*, March 8, 2017.

Israel Aerospace Industries, "Heron," undated (a). As of August 21, 2017:
http://www.iai.co.il/2013/18900-16382-en/BusinessAreas_UnmannedAirSystems_HeronFamily.aspx

———, "Heron Family," undated (b). As of December 20, 2017:
http://www.iai.co.il/2013/18900-en/BusinessAreas_UnmannedAirSystems_HeronFamily.aspx

Joint Chiefs of Staff, *Close Air Support*, Joint Publication 3-09.3, November 25, 2014. As of August 15, 2017:
https://fas.org/irp/doddir/dod/jp3_09_3.pdf

"Kazakhstan to Start UAV Assembly in 2017," *UAS Vision*, September 26, 2016.

Khan, Bilal, "Could the CASC CH-5 UAV Be an Option for Pakistan?" Quwa Defence News and Analysis Group, July 20, 2017. As of October 3, 2017:
http://quwa.org/2017/07/20/pakistan-consider-casc-ch-5-uav/

Kreps, Sarah, "Drone Proliferation: What We Have to Fear," *Hill*, June 25, 2014. As of August 3, 2017:
http://thehill.com/blogs/pundits-blog/210109-drone-proliferation-what-we-have-to-fear

Lambeth, Benjamin S., *Air Operations in Israel's War Against Hezbollah: Learning from Lebanon and Getting It Right in Gaza*, Santa Monica, Calif.: RAND Corporation, MG-835-AF, 2011. As of December 14, 2017:
https://www.rand.org/pubs/monographs/MG835.html

Lappin, Yaakov, "Air Platforms: IAI Pitches MTCR-Compliant Heron TP to India," *Jane's Defence Weekly*, February 10, 2017. As of December 20, 2017:
http://www.janes.com/article/67623/iai-pitches-mtcr-compliant-heron-tp-to-india

Lennane, Alex, "Unmanned Aircraft Are the Future for Air Cargo: 'But We Need Time to Get It Right,'" *Loadstar*, May 26, 2015. As of September 19, 2017:
https://theloadstar.co.uk/unmanned-aircraft-future-air-cargo-need-time-get-right/

Leonardo Company, "Falco EVO," undated. As of October 3, 2017:
http://www.leonardocompany.com/en/-/falco-evo

Lin, Jeffrey, and P. W. Singer, "Meet China's Sharp Sword, a Stealth Drone That Can Likely Carry 2 Tons of Bombs," *Popular Science*, January 18, 2017a. As of August 29, 2017:
http://www.popsci.com/china-sharp-sword-lijian-stealth-drone

———, "In China, an E-Commerce Giant Builds the World's Biggest Delivery Drone," *Popular Science*, May 24, 2017b. As of September 8, 2017:
http://www.popsci.com/jd-com-builds-worlds-biggest-delivery-drone

Lockheed Martin, "Universal Ground Control Station (UGCS)," undated. As of August 10, 2017:
http://www.lockheedmartin.com/us/products/cdl-systems/about-us/projects/universal-ground-control-station.html

Marques, Mário Monteiro, *Standard Interfaces of UAV Control System (UCS) for NATO UAV Interoperability*, North Atlantic Treaty Organization, Science and Technology Organization, STO-EN-SCI-271, Standardization Agreement 4586, May 1, 2015. As of December 14, 2017:
https://www.sto.nato.int/publications/STO%20Educational%20Notes/STO-EN-SCI-271/EN-SCI-271-03.pdf

Mayer, John E., Program Executive Office, Unmanned Aviation and Strike Weapons, and author of *State of the Art of Airworthiness Certification* (North Atlantic Treaty Organization, Science and Technology Organization, STO-MP-AVT-273, undated), interview with author, August 1, 2017.

Meredith, Sam, and Arjun Kharpal, "Chinese E-Commerce Giant JD.com Is Developing a Drone That Can Deliver Packages Weighing as Much as One Ton," CNBC, June 8, 2017. As of September 12, 2017:
https://www.cnbc.com/2017/06/08/e-commerce-jdcom-alibaba-amazon-drone-delivery-china-asia-technology.html

Missile Technology Control Regime, homepage, undated (a). As of August 14, 2017:
http://www.mtcr.info

———, "Frequently Asked Questions (FAQs)," undated (b). As of September 4, 2017:
http://mtcr.info/frequently-asked-questions-faqs/

———, "Guidelines for Sensitive Missile-Relevant Transfers," undated (c). As of November 13, 2017:
http://mtcr.info/guidelines-for-sensitive-missile-relevant-transfers/

———, "MTCR Partners," undated (d). As of October 22, 2017:
http://mtcr.info/partners/

———, *Missile Technology Control Regime (MTCR) Annex Handbook*, 2010. As of August 21, 2017:
http://mtcr.info/wordpress/wp-content/uploads/2016/04/MTCR_Annex_Handbook_ENG.pdf

MTCR—*See* Missile Technology Control Regime.

Muralidharan, Rathna K., "Trump's Saudi Arms Deal: A Historical Boost for U.S. Industry," *RealClearDefense*, August 24, 2017. As of September 17, 2017:
https://www.realcleardefense.com/articles/2017/08/24/trumps_saudi_arms_deal__a_historical_boost_for_us_industry_112131.html

NATO—*See* North Atlantic Treaty Organization.

North Atlantic Treaty Organization, *(Restricted) Interoperable Data Links for Imaging Systems*, Standardization Agreement 7085, January 15, 2004.

Northrop Grumman, "MQ-4C Triton Improves Mission Capability with Successful Software Upgrade Test," undated (a). As of September 1, 2017:
http://news.northropgrumman.com/file?fid=58e29712a138353201a86190

———, "MQ-4C Triton: Making the World's Oceans Smaller," undated (b). As of August 16, 2017:
http://www.northropgrumman.com/Capabilities/Triton/Pages/default.aspx

———, "MQ-5B Hunter," 2012. As of October 3, 2017:
http://www.northropgrumman.com/Capabilities/MQ5BHunter/Documents/Hunter_Data_Sheet.pdf

Office of the Secretary of Defense, *Unmanned Aircraft Systems Roadmap, 2005–2030*, Washington, D.C., August 4, 2005. As of August 15, 2017:
https://fas.org/irp/program/collect/uav_roadmap2005.pdf

Page, Jeremy, and Paul Sonne, "Unable to Buy U.S. Military Drones, Allies Place Orders with China," *Wall Street Journal*, July 17, 2017.

Piaggio Aerospace, "P.1HH HammerHead," undated. As of September 7, 2017:
http://www.p1hh.piaggioaerospace.it/

"PM Narendra Modi's Israel Visit: India Likely to Get Heron-TP Armed Drones," *Indian Express*, July 4, 2017. As of October 3, 2017:
http://indianexpress.com/article/india/
pm-narendra-modis-israel-visit-india-likely-to-get-heron-tp-armed-drones-4735276/

Pomerleau, Mark, "Future of Unmanned Capabilities: MALE vs HALE," *Defense Systems*, May 27, 2015. As of August 14, 2017:
https://defensesystems.com/articles/2015/05/27/uas-male-vs-hale-debate.aspx

Public Law 114-328, National Defense Authorization Act for Fiscal Year 2017, December 23, 2016. As of December 15, 2017:
https://www.gpo.gov/fdsys/pkg/PLAW-114publ328/content-detail.html

Rawnsley, Adam, "Ukraine Scrambles for UAVs, but Russian Drones Own the Skies," *War Is Boring*, February 20, 2015. As of September 23, 2017:
https://warisboring.com/ukraine-scrambles-for-uavs-but-russian-drones-own-the-skies/

———, "Meet China's Killer Drones," *Foreign Policy*, January 14, 2016. As of August 21, 2017:
http://foreignpolicy.com/2016/01/14/meet-chinas-killer-drones/

"Raytheon to Equip GA-ASI's MQ-9 Reaper UAS with MALD," *Air Force Technology*, February 17, 2013. As of September 13, 2017:
http://www.airforce-technology.com/news/newsraytheon-equip-general-atomics-reaper-uas-mald

Reed, John, "Meet China's New-Old Killer Drones," *Foreign Policy*, January 8, 2013. As of December 15, 2017:
http://foreignpolicy.com/2013/01/08/meet-chinas-new-old-killer-drones/

Reiner Stemme Utility Air Systems, "Q01: Mission-Specific MALE Platform—Features," undated (a). As of December 22, 2017:
http://www.rs-uas.com/products/q01-airborne-platform/features

———, "Q01: Mission-Specific MALE Platform—Technical Specs," undated (b). As of December 16, 2017:
http://www.rs-uas.com/products/q01-airborne-platform/technical-specs/

Ross, Alice, "UK Drones Three Times More Likely Than US to Fire in Afghanistan," Bureau of Investigative Journalism, September 6, 2013. As of October 3, 2017:
https://www.thebureauinvestigates.com/stories/2013-09-06/
uk-drones-three-times-more-likely-than-us-to-fire-in-afghanistan

Rubin, Uzi, "Israel's Defence Industries: An Overview," *Defence Studies*, Vol. 17, No. 3, 2017, pp. 228–241.

Sanders, Ralph, "An Israeli Military Innovation: UAVs," *Joint Force Quarterly*, Winter 2002–2003, pp. 114–118. As of December 15, 2017:
http://oai.dtic.mil/oai/oai?verb=getRecord&metadataPrefix=html&identifier=ADA483682

"Saudi Arabia Imports UAV Production Line from China: Reports," *People's Daily Online*, March 27, 2017. As of September 18, 2017:
http://en.people.cn/n3/2017/0327/c90000-9195600.html

Sayler, Kelley, Ben FitzGerald, Michael C. Horowitz, and Paul Scharre, *Global Perspectives: A Drone Saturated Future*, Washington, D.C.: Center for a New American Security, May 2016. As of August 4, 2017:
http://drones.cnas.org/reports/global-perspectives/

Scarborough, Rowan, "Obama Rejected Jordanian King Abdullah's Pleas for Predator Drones," *Washington Times*, February 5, 2015. As of August 15, 2017:
http://www.washingtontimes.com/news/2015/feb/5/obama-denied-request-jordan-predator-drones-islami/

Schulberg, Jessica, "Why Is the U.S. So Stingy with Its Drones? It's Costing Us," *New Republic*, July 2, 2014. As of August 10, 2017:
https://newrepublic.com/article/118528/americas-drone-export-policy-costing-us-money-and-influence

Security Assistance Management Manual, Defense Security Cooperation Agency, "Welcome to DSCA's E-SAMM and Policy Memoranda Distribution Portal," Washington, D.C., undated. As of August 21, 2017:
http://www.samm.dsca.mil/

Sharp, Jeremy M., *U.S. Foreign Aid to Israel*, Washington, D.C.: Institute for Palestine Studies, March 12, 2012. As of December 15, 2017:
http://www.palestine-studies.org/books/us-foreign-aid-israel-march-12-2012#about-author

SIPRI—*See* Stockholm International Peace Research Institute.

Stockholm International Peace Research Institute, "SIPRI Arms Transfers Database," undated (a). As of September 14, 2017:
https://sipri.org/databases/armstransfers

———, "SIPRI Databases," undated (b). As of September 14, 2017:
https://www.sipri.org/databases

Stohl, Rachel, *UAV Export Controls and Regulatory Challenges: Working Group Report*, Washington, D.C.: Stimson Center, October 13, 2015. As of December 15, 2017:
https://www.stimson.org/content/export-controls-and-regulatory-challenges

Stott, Michael, "Deadly New Russian Weapon Hides in Shipping Container," Reuters, April 26, 2010. As of October 3, 2017:
http://www.reuters.com/article/us-russia-weapon-idUSTRE63P2XB20100426

Teal Group, "Teal Group Predicts Worldwide Civil Drone Production Will Soar $73.5 Billion over the Next Decade," press release, June 19, 2017. As of December 15, 2017:
http://www.tealgroup.com/index.php/about-teal-group-corporation/press-releases/136-teal-group-predicts-worldwide-civil-drone-production-will-soar-73-5-billion-over-the-next-decade

Thomson, Iain, "Chinese E-Tailer Beats Amazon to the Skies with One-Ton Delivery Drones," *Register*, May 23, 2017. As of September 19, 2017:
https://www.theregister.co.uk/2017/05/23/chinese_etailer_drones_beat_amazon/

"UK Drone Strike Stats," *Drone Wars UK*, updated August 2017. As of September 14, 2017:
https://dronewars.net/uk-drone-strike-list-2/

"UK Team Working to Develop Mid-Mass Logistics Drone," *Inside Unmanned Systems*, undated. As of September 19, 2017:
http://insideunmannedsystems.com/uk-team-working-develop-mid-mass-logistics-drone/

"Unblinking Eyes in the Sky," *Economist*, March 3, 2012. As of December 15, 2017:
http://www.economist.com/node/21548485

"United Arab Emirates: Air Force," *Jane's World Air Forces*, July 2, 2017.

"United Kingdom: Air Force," *Jane's World Air Forces*, July 11, 2017.

United Nations Security Council, Libya Sanctions Committee, "Final Report of the Panel of Experts in Accordance with Paragraph 13 or Resolution 2278 (2016)," United Nations, June 1, 2017.

Unmanned aerial vehicle industry representatives, interview with the authors, Washington, D.C., July 26, 2017a (name withheld by agreement).

Unmanned aerial vehicle industry representatives, interview with the authors, California, August 8, 2017b (name withheld by agreement).

U.S. Air Force, "MQ-9 Reaper," September 23, 2015. As of August 21, 2017:
http://www.af.mil/About-Us/Fact-Sheets/Display/Article/104470/mq-9-reaper/

"US Approves Sale of 22 SkyGuardians to India," *UAS Vision*, June 26, 2017. As of October 3, 2017:
http://www.uasvision.com/2017/06/26/us-approves-sale-of-22-skyguardians-to-india/

U.S. Department of Commerce, Bureau of International Security, "MTCR Controls and UAVs," presentation to Association of Unmanned Vehicle Systems International, Dallas, Texas, June 2017.

U.S. Department of State, "U.S. Export Policy for Military Unmanned Aerial Systems," Washington, D.C., February 17, 2015. As of August 21, 2017:
https://2009-2017.state.gov/r/pa/prs/ps/2015/02/237541.htm

U.S. Government Accountability Office, *Nonproliferation: Agencies Could Improve Information Sharing and End-Use Monitoring on Unmanned Aerial Vehicle Exports*, Washington, D.C., GAO-12-536, July 30, 2012. As of December 15, 2017:
http://www.gao.gov/products/GAO-12-536

U.S. Marine Corps, *2017 Marine Aviation Plan*, undated. As of December 15, 2017:
http://www.aviation.marines.mil/Portals/11/2017%20MARINE%20AVIATIOIN%20PLAN.pdf

Vergun, David, "Apache-UAV Teaming Combines 'Best Capabilities of Man, Machine,'" U.S. Army, May 8, 2014. As of October 3, 2017:
https://www.army.mil/article/125676

Vick, Alan J., *Air Base Attacks and Defensive Counters: Historical Lessons and Future Challenges*, Santa Monica, Calif.: RAND Corporation, RR-968-AF, 2015. As of December 15, 2017:
https://www.rand.org/pubs/research_reports/RR968.html

Wheeler, Winslow, "The MQ-9's Cost and Performance," *Time*, February 28, 2012. As of August 21, 2017:
http://nation.time.com/2012/02/28/2-the-mq-9s-cost-and-performance/

Williams, Huw, "IDEX 2017: General Atomics to Offer Predator XP as COCO Capability," *Jane's International Defence Review*, February 22, 2017. As of October 1, 2017:
http://www.janes.com/article/68094/idex-2017-general-atomics-to-offer-predator-xp-as-coco-capability

Wong, Kelvin, "Air Platforms: Heavily Armed CASC CH-5 UAV Makes Public Debut," *IHS Jane's International Defence Review*, November 7, 2016.

"X-47B Unmanned Combat Air System (UCAS)," *Naval Technology*, undated. As of September 6, 2017:
http://www.naval-technology.com/projects/x-47b-unmanned-combat-air-system-carrier-ucas/

Yeo, Mike, "China Deploys New Anti-Submarine Aircraft to Fringes of South China Sea," *Military Times*, June 22, 2017. As of October 3, 2017:
https://www.militarytimes.com/space/2017/06/22/china-deploys-new-anti-submarine-aircraft-to-fringes-of-south-china-sea/